Second Edition

Basic Chemistry for Biology

Carolyn Chapman
Suffolk County Community College

Boston Burr Ridge, IL Dubuque, IA Madison, WI New York San Francisco St. Louis
Bangkok Bogotá Caracas Lisbon London Madrid
Mexico City Milan New Delhi Seoul Singapore Sydney Taipei Toronto

WCB/McGraw-Hill

A Division of The **McGraw-Hill** *Companies*

BASIC CHEMISTRY FOR BIOLOGY, SECOND EDITION

This book is printed on acid-free paper.

1 2 3 4 5 6 7 8 9 0 QPF/QPF 9 3 2 1 0 9 8

ISBN 0-697-36087-3

Vice president and editorial director: *Kevin T. Kane*
Publisher: *Michael D. Lange*
Sponsoring editor: *Patrick Reidy*
Senior developmental editor: *Suzanne M. Guinn*
Marketing manager: *Lisa Gottschalk*
Project manager: *Jill R. Peter*
Senior production supervisor: *Sandra Hahn*
Designer: *K. Wayne Harms*
Compositor: *Shepherd, Inc.*
Typeface: *10/12 Times Roman*
Printer: *Quebecor Printing Book Group/Fairfield, PA*

Cover/interior design: *Kristyn Kalnes*
Cover photograph: *Stephen Simpson/FPG International*

www.mhhe.com

Table of Contents

Introduction

Welcome to the second edition of *Basic Chemistry for Biology.* Chemistry is a subject that is important in many other areas of science. Students of biology, medicine, geology, and astronomy need a working knowledge of chemistry to succeed in their fields. The goal of this book is to provide you with an introduction or review of the basic principles of chemistry that are most useful in other areas of science.

As a beginning science student, you may wonder why you need to know about chemistry if your direct goal is to learn a field such as anatomy and physiology or geology. Some examples will illustrate why chemistry is so fundamental. The human body, like that of all living organisms, is composed of chemicals. The composition of the body will not make sense unless a basic knowledge of chemistry is available to help you understand the body's structure. Living organisms are based on chemical activity. When the liver responds to hormones and regulates blood sugar levels, chemical reactions are responsible. When you think and learn, chemical interactions in the neurons of the brain play an essential role. These examples show that physiology is the study of how the body regulates its internal chemistry. Clearly, a knowledge of chemistry will greatly increase your chances of success in physiology. Similar examples from other areas of science would lead to the same conclusion.

This book can be used as an introduction to chemistry or as a review of the subject. Those topics from high school chemistry (or an entry-level college course) that are most essential to an understanding of other science areas are included here. The second edition includes new chapters on organic chemistry and biomolecules. Although these topics are not usually covered in high school chemistry, they are essential to your success in biology. This book is not intended to replace an entire chemistry course. If you master the subject matter of this book, you will learn (or relearn) the principles of chemistry used in introductory science courses. If you decide to pursue the study of science seriously, however, you will need to take several chemistry courses as you progress.

All fields of science are based on experimentation. The goal of this book is to help you acquire a basic knowledge of the concepts of chemistry as quickly as possible. Therefore, the concepts of chemistry are presented without their experimental basis. It is important to remember that these concepts are not arbitrary but were developed to explain and be compatible with experimental findings. A full chemistry course will include the experimental basis of the field.

A scientific model is a description and explanation of experimental observations. Models can vary in complexity. In general, simpler models are utilized in this book. The octet rule, used to predict chemical reactions, is an example. Such models are compatible with many experimental findings and can be used to predict a wide array of chemical reactions. Typically, they do not explain everything, and exceptions are known. This does not mean that such models are incorrect, but rather that they represent incomplete and simplified descriptions of the natural world. As your knowledge of chemistry increases, more complex models can be added to those presented here.

How To Use This Book

Objectives

Each chapter in this book covers the major topic stated in its title. The first page of each chapter gives a list of learning objectives for the chapter. These objectives list the things you should be able to do when you have mastered the material in the chapter.

Some of these objectives are basic and must be mastered before you can continue on to more advanced topics. This is because much of the subject matter of chemistry is cumulative, and later topics build on what has gone before. Other objectives are intermediate or advanced. These objectives are indicated with an asterisk. Frequently, these topics can be omitted without impeding your progress through future chapters. You should customize your chemistry study by choosing which of these to include. Your prior knowledge of chemistry and the requirements of your current academic program are prime factors to consider. One effective strategy is to work through the book two or more times. Cover the basic objectives the first time. This provides a good start to the beginning student without introducing difficult subject matter. Then, during subsequent readings, add intermediate and advanced topics.

Study and Self-Testing

Each section of text includes a short paragraph about a new topic. Read the paragraph carefully. The paragraph will be followed by a series of questions that test your understanding of the material and ask you to apply it. Answer the questions as completely as you can.

The answers to the questions are at the end of the chapter. When you have completed the questions on a page, check your responses against the answers. Explanations are often included with the answers. If you did not answer the question correctly, the explanation may be sufficient to help you get back on track.

If you understood the paragraph and answered all the questions correctly, you are ready to continue to the next section. If you are uncertain that you understand or if you missed questions, repeat the section. Read the paragraph again. When you understand it, try the questions again.

Repetition is essential to the learning process. Working through this book more than once provides such repetition and is strongly recommended. A longer and more complete test is included at the end of each chapter. This test enables you to evaluate your mastery of the entire chapter. The chapter test is also a learning experience, because it provides more repetition of the subjects that have been covered.

Chapter 1

The Atom

How to Use the Objectives

The objectives for Chapter 1 are stated below. All are tasks you should be able to complete successfully once the material in this chapter has been mastered. These are general objectives. The questions in Chapter 1 specifically apply to these objectives. Correct answers to the questions in Chapter 1 provide good evidence that the objectives have been achieved. A full chapter test at the end of this unit is available for you to take when you are ready. It includes questions designed to evaluate all of the objectives. Your score on the chapter test provides you with an overall assessment of your mastery of the chapter objectives.

Objectives that cover more advanced material are indicated with an asterisk (*). Include the optional objectives appropriate for your academic program. Progress through the remaining chapters of this book does not require mastery of the optional topics.

Objectives

1. Define the following terms and apply the definitions correctly: matter, element, atom, atomic number, mass number, atomic mass (weight), and isotope.
2. List the elementary particles, their properties, and their locations in an atom.
3. State the interactions of like and unlike electrical charges.
4. Given the atomic number and mass number, calculate the number of protons, neutrons, and electrons in an atom.
5. Use the periodic table of the elements to find information such as an element's name, atomic number, atomic mass (weight), and symbol.
6. List the symbols for the elements that are significant in living things.
7. Know and use the appropriate formula to calculate the number of electrons in an electron shell.
8. Sketch the structure of any atom up to an atomic number of 18, showing the correct number of electrons in each electron shell.
9. Represent the isotopic variants of an element with standard chemical notation.
10. *Sketch the electronic structure of any atom up to atomic number 38, showing the correct number of electrons in each electron shell.
11. *Accurately describe the size of an atom and the spatial relations within it.

Matter, Elements, and Atoms

All physical objects are made of matter. Your body, all the objects you see, and even the air that fills the space between visible objects are composed of matter. (Matter in a solid or liquid state is usually visible, whereas matter in the gaseous state often is not.) Matter is defined as anything that occupies space and has mass. The space an object occupies can be measured and is the volume of that object. Under the circumstances we usually encounter (on earth in a gravitational field), having mass is detected as the weight of an object.

Questions

1. Is your body composed of matter? How would you demonstrate the validity of your answer?

   Yes ______ No ______

2. Is the chair or couch you are sitting on composed of matter? How can you tell?

   Yes ______ No ______

3. A tank appears to be empty. It is weighed. Then it is connected to a vacuum pump, which is operated for ten minutes. The tank is weighed again and now weighs less. Did it contain matter? How did the information provided permit you to answer the question?

   Yes ______ No ______

How could you determine whether the tank still contains some matter after being pumped?

Chemists have analyzed matter carefully. Their experimental work shows that matter is made up of simple substances called elements. The elements are the basic building blocks of more complex forms of matter. Elements cannot be converted to less complex substances by chemical reactions.

Questions

4. Samples of four pure gases are put through testing procedures that attempt to decompose them into simpler substances using chemical reactions. The following results are obtained: oxygen, nitrogen, and neon cannot be decomposed into simpler substances. Carbon dioxide does decompose, releasing carbon and oxygen. Based on these results, which gases are elements? Explain.

5. It is possible to use an electrical current to decompose a sample of water, a process called electrolysis. As the electrolysis procedure progresses, both hydrogen gas and oxygen gas are released. Is water an element? What evidence supports your answer?

   Yes ______ No ______

6. A pure sample of sugar is burned in oxygen. The products of combustion are carbon dioxide and water. Is sugar an element? Explain your reasons for answering as you did.

   Yes ______ No ______

Experimental investigation demonstrates that matter is particulate. An atom is the smallest particle of an element that still retains the properties of that element. Atoms are extremely tiny. Between one million and two million atoms could line up side by side across the diameter of one of the printed periods on this page.

Questions

7. Carbon is an element that is very important in organic molecules. (Organic molecules are synthesized by living cells and are always based on carbon. Organic molecules are central to the structure and function of all forms of life.) An atom of carbon is one ______ of carbon.

8. Experimentation shows that atoms can be broken up, though not by chemical means. Atoms have internal structure and contain still smaller particles. Would the smaller particles within the carbon atom have the properties of carbon? Explain.

   Yes ______ No ______

9. Would one particle of sugar be an atom? Explain your answer.

   Yes ______ No ______

10. The smallest particle of nitrogen that still has the properties of nitrogen is one ______ of nitrogen. Would this single particle of nitrogen be visible to the unaided eye? Explain your answer.

    Yes ______ No ______

Elementary Particles

The simplified model of atomic structure presented here considers the atom to be composed of three types of elementary particles (subatomic particles): the proton, the neutron, and the electron. Each particle is characterized by a weight (or mass) and an electrical charge.

Particle	Symbol	Mass*	Charge
Proton	p	1	+1
Neutron	n	1	0
Electron	e^-	0**	−1

*The units of mass are atomic mass units (amu) or daltons.
**The mass of the electron is not exactly zero, but is so small that it is negligible for most purposes.

Questions

11. Which elementary particles have a charge?

12. Which elementary particle is not charged?

13. Which elementary particles account for nearly all the weight of the atom?

14. Which elementary particle has a charge that is equal in magnitude, but opposite in sign, to the proton?

15. Which elementary particle contributes very little to the weight of the atom?

Electrical charges that are opposite in sign attract each other. Electrical charges that have the same sign repel each other. Uncharged matter neither repels nor attracts other matter.

Questions

For the pairs of particles listed, state whether the particles attract each other, repel each other, or do not interact.

16. Proton and neutron.

17. Proton and electron.

18. Electron and neutron.

19. Neutron and neutron.

20. Electron and electron.

21. Proton and proton.

The Atom

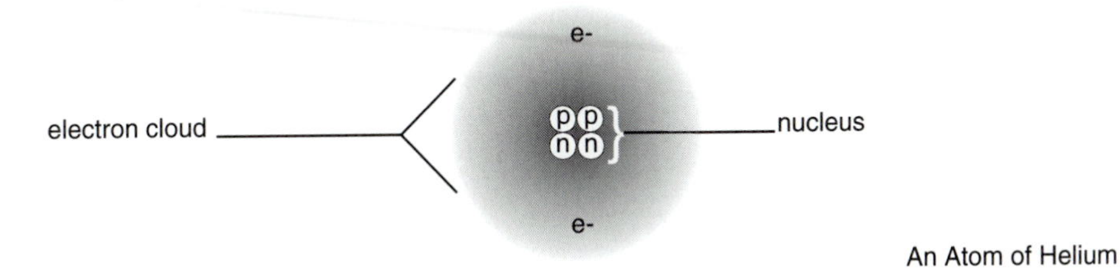

An Atom of Helium

This diagram of an atom* shows the locations of the elementary particles. The nucleus of the atom contains the protons (p) and neutrons (n). The nucleus is extremely small and therefore very dense. The electrons (e-) are located at a distance from the nucleus. The electron cloud represents the regions where the electrons are most likely to be found. The electrons in the cloud may be pictured as moving very rapidly in the area of the cloud, or as a smear of negative charge.

*This diagram represents an atom of helium. The helium atom has two protons, two electrons, and two neutrons. The numbers of elementary particles would be different in atoms of other elements.

Questions

22. The atom is mostly empty space. Explain your answer.

    True ______ False ______

23. Which part of the atom is positively charged?

24. Most of the mass of the atom is contained in the ___________.

25. Sketch an atom that contains three protons, three electrons, and four neutrons.

The atomic number of an element is equal to the number of protons the element contains. The atomic number of an element establishes its identity. For example, hydrogen is atomic number 1, carbon is atomic number 6, and nitrogen is atomic number 7.

Questions

26. How many protons does carbon contain?

27. Where are the protons located in the carbon atom?

28. Could an atom with an atomic number of 8 be a variant of nitrogen? Explain.

    Yes ______ No ______

29. How many protons do hydrogen atoms contain?

30. How many protons do nitrogen atoms contain?

Unreacted atoms are electrically neutral. Because they contain charged elementary particles, the number of positive and negative charges must be the same.

Questions

31. Which elementary particle is positively charged?

32. Which elementary particle is negatively charged?

33. In a neutral atom, the number of protons must be equal to the number of ________________. Explain.

34. An atom of carbon contains ______ protons and ______ electrons. Explain how you determined the answer to this question.

35. An atom of nitrogen contains ______ protons and ______ electrons.

36. How many electrons does a hydrogen atom contain? Explain.

The Periodic Table of the Elements

The elements are arranged in a sequence by increasing atomic number in a chart called the periodic table of the elements. The periodic table organizes a great deal of information about the elements in a very compact form. You will learn to use some of that information now. A copy of the periodic table (or chart) is located at the back of this book. The key at the top of the periodic table explains the information that is available about each element. Use the key and periodic table to answer the following questions.

Questions

37. Where is the atomic number for an element found?

38. Where is the name of the element found?

39. The atomic number of calcium is ______.

40. The atomic number of sodium is ______.

41. Which element has an atomic number of 26?

42. Which element has an atomic number of 12?

Each element is represented by a chemical symbol. The symbol is often the first letter or two of the element's name in English. Occasionally, the symbol is derived from the name of the element in Latin. Use the periodic table to answer the following questions.

Questions

43. Carbon, hydrogen, oxygen, and nitrogen are four elements that are present in substantial amounts in all living organisms. What are the symbols for these four elements?

44. Look up the atomic numbers and chemical symbols for phosphorus and potassium.

45. What are the chemical symbols for sodium, sulfur, and iron?

46. Which element has an atomic number of 17? What is the chemical symbol for this element? How many protons does an atom of this element contain? How many electrons?

The table below lists many of the elements that are biologically important. The biological role of the element is summarized.

Questions

47. Complete the table by looking up the symbols and atomic numbers in the periodic table of the elements.

Element	Symbol	Atomic #	Biological Roles
Hydrogen			Present in almost all organic molecules
Carbon			Forms the backbone of all organic molecules
Nitrogen			Found in all amino acids, proteins, and nucleic acids
Oxygen			Present in varying amounts in many organic molecules
			Essential gas for aerobic metabolism
Sodium			Essential in tissue fluids and nerve impulse conduction
Magnesium			Required for many reactions in cells
Phosphorus			Found in all nucleic acids, some lipids, and adenosine tri-phosphate (ATP)
Sulfur			Present in some amino acids
Chlorine			Present in tissue fluids
Potassium			Essential in tissue fluids and nerve impulse conduction
Calcium			Involved in nerve impulse transmission and muscle contraction, and strengthens bone
Iron			Present in hemoglobin, which transports oxygen
Zinc			Required for many reactions in cells

Isotopes

Every atom is characterized by two numbers. The atomic number, equal to the number of protons, has already been introduced. The second number is called the mass number. The mass number is the sum of the protons and neutrons in the nucleus.

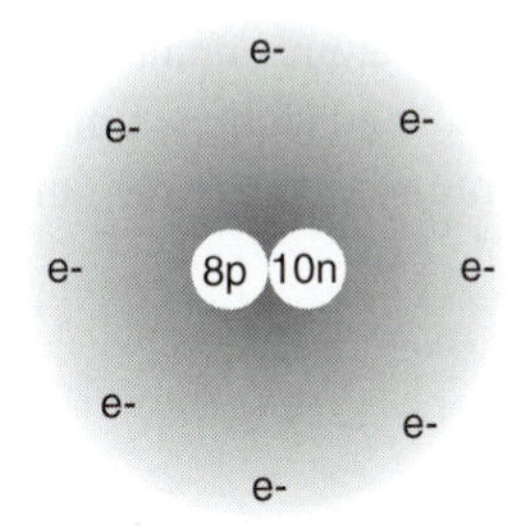

Questions

48. The atomic number for the element diagrammed above is ______.

    The mass number for the element diagrammed above is ______.

49. The atom contains ______ neutrons.

50. Devise a formula for calculating the number of neutrons in an atom when a diagram is not available. Assume you do know the atomic number and mass number.

51. In the space below, sketch an atom with an atomic number of 8 and a mass number of 16. Compare your atom to the one diagrammed above, and explain how they are the same and how they differ.

Atoms having the same atomic number but different mass numbers are isotopes. The isotopic variants of a given element generally show the same chemical behavior. Some isotopes are physically stable. Other isotopes are unstable and undergo radioactive decay.

Questions

52. In the space below, sketch an atom with an atomic number of 6 and a mass number of 12. Label this sketch "Atom 1." Sketch another atom with an atomic number of 6 and a mass number of 13. Label this sketch "Atom 2." Sketch a third atom with an atomic number of 6 and a mass number of 14. Label this sketch "Atom 3."

53. Atom 1 is the element ____________. This atom contains ______ neutrons in its ____________.

54. Atom 2 is the element ____________. This atom contains ______ neutrons. Atom 3 is the element ____________. This atom contains ______ neutrons.

55. Atoms 1, 2, and 3 are ____________ of each other.

56. Isotopic variants of a given element contain the ______ number of protons, the ______ number of electrons, and ____________ numbers of neutrons.

57. Radioactive isotopes (also called radioisotopes) are ____________.

58. Variants of an element that have different numbers of neutrons are called ____________.

Chemical notation is very compact. An isotope of an element can be represented as:

$$^{A}_{Z}X$$

In this notation, X represents the chemical symbol for the element, Z is the atomic number for the element, and A is the mass number for the isotope under consideration.

Questions

59. Using the notation presented above, represent the three isotopic variants of carbon diagrammed below.

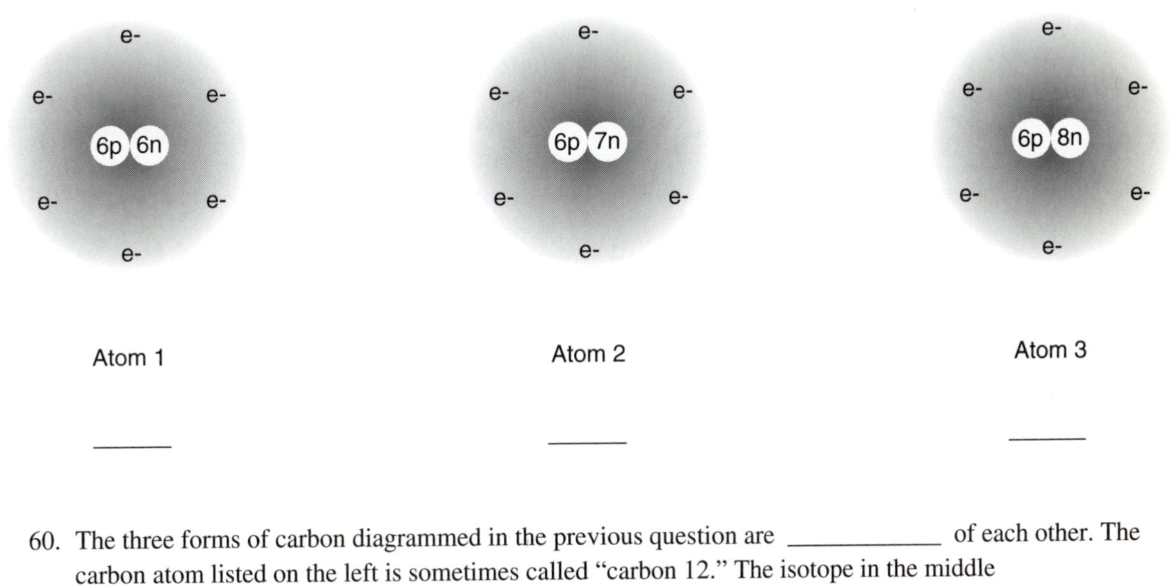

______ ______ ______

60. The three forms of carbon diagrammed in the previous question are ____________ of each other. The carbon atom listed on the left is sometimes called "carbon 12." The isotope in the middle is ____________ and the isotope on the right is ____________.

61. In the space below, sketch an atom of $^{6}_{3}Li$. (Li is the chemical symbol for lithium.)

This isotopic variant of lithium is called ____________.

The atomic mass (also called the atomic weight) of an element is the weighted average of the various isotopes of the element that are found in nature. For certain elements, only one prevalent isotopic form exists. If nearly all the atoms of an element are the same isotope, the atomic weight will be almost the same as the mass number of that isotope. In such cases, the atomic weight will be close to a whole number. Atomic weights are obtained from the periodic table of the elements. Use your periodic table to answer the following questions.

Questions

62. What are the atomic weights for hydrogen, carbon, oxygen, magnesium, and copper?

63. Which of the elements from question 62 have atomic weights that are close to whole-number values?

64. Which of the elements from question 62 are likely to have only one common isotopic form in a typical sample of that element? Explain the reasons for your answer.

65. Fill in the table below. Give the atomic number of the element and the mass number of the most common isotope, and calculate the number of neutrons present in that isotope. Explain how the number of neutrons is calculated.

Element	Atomic Number	Mass Number	# of Neutrons
Hydrogen			
Carbon			
Oxygen			

*Intermediate Difficulty

For some elements, two or more isotopic variants occur naturally. A sample of such an element will contain atoms of the different isotopic types in the proportion that they occur in nature. Atoms of the different isotopes will have different weights and thus different mass numbers. If two or more isotopes are prevalent for an element, the atomic weight will be between the mass numbers of the common isotopes and may not be close to a whole number.

*Questions

The questions with an asterisk are intermediate (*) to advanced (**) in difficulty.

*Intermediate Difficulty Sections: Sections marked intermediate or advanced may be omitted by the beginning student. Future sections do not depend on this material. You can customize your chemistry study by including only the more advanced topics that you need to know for your academic program.

66. Of the elements hydrogen, carbon, oxygen, magnesium, and copper, which have atomic weights that are not close to whole-number values? Explain.

67. There are two common isotopes of copper with mass numbers of 63 and 65. For each of these isotopes, calculate the number of protons, electrons, and neutrons an atom of that isotope would contain. Explain how these calculations are carried out.

Isotope	# Protons	# Electrons	# Neutrons
$^{63}_{29}Cu$			
$^{65}_{29}Cu$			

68.*Which isotope of copper is most common in natural samples? To answer this question, compare the atomic weight of copper to the mass numbers of the two isotopes. Explain your reasoning.

69.**Do any individual copper atoms weigh 63.54 amu (atomic mass units)? Explain.

Yes ______ No ______

Electronic Structure of Atoms

The electronic structure of an atom is the best predictor of its chemical behavior. A simplified model of the electronic structure of atoms is presented in this section and sketched in the diagram below. In this model, the electrons are located in principal energy levels (these are also referred to as electron shells). Each principal energy level can contain a different number of electrons. The maximum capacity of a principal energy level (electron shell) is given by the formula $2n^2$, where n stands for the principal energy level.

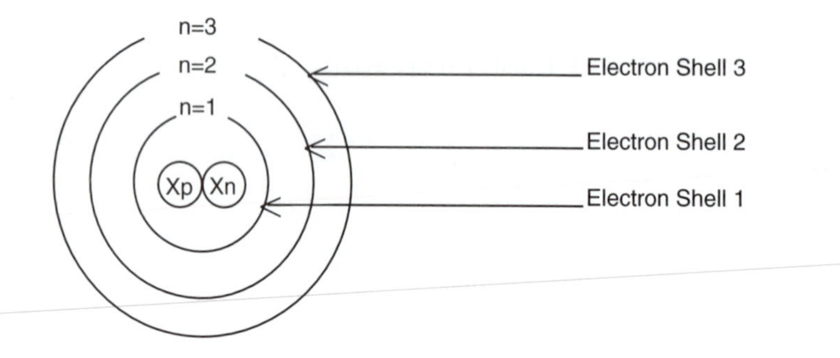

Questions

70. The first electron shell, principal energy level one, can contain a maximum of ______ electrons. Could this shell contain one electron?

    Yes ______ No ______

    Could this shell contain four electrons? Explain.

    Yes ______ No ______

71. When n = 2, the maximum number of electrons in the shell is ______. Could principal energy level two contain seven electrons?

    Yes ______ No ______

    Could it contain eleven electrons?

    Yes ______ No ______

72. The maximum number of electrons in the third electron shell (n = 3) is ______. Could this energy level contain twenty-six electrons?

    Yes ______ No ______

When n = 1, the electrons are closest to the nucleus and have a lower energy level. In general, as n increases, electrons are located farther from the nucleus and contain greater energy. Electrons typically occupy the lowest energy level available. Although an electron cloud is a more accurate way of visualizing electrons, for convenience, the electron shells are sketched as circles. The circles represent a distance from the nucleus where the electrons are likely to be found, but do not indicate exact positions or shapes.

Questions

For each of the elements listed below, look up the atomic number, determine how many electrons an atom contains, determine how many electrons are present in each electron shell, and sketch an atom. The first question, the element carbon, is worked as an example.

73. Carbon

    Atomic # = 6

    # electrons = 6 (see below for explanation•)

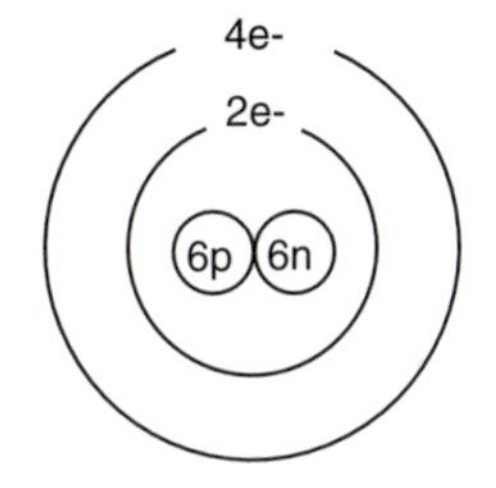

74. Chlorine

    Atomic # = ______

    # electrons = ______

75. Hydrogen

    Atomic # = ______

    # electrons = ______

76. Sodium

    Atomic # = ______

    # electrons = ______

•Two electrons fill the first electron shell. The remaining four electrons occupy the second electron shell.

For the first eighteen elements, the electrons fill the principal energy levels (electron shells), as predicted by simple rules. That is, electrons occupy the lowest electron shell that is not filled to capacity. The formula $2n^2$ is used to calculate the maximum capacity of each electron shell.

Questions

77. The table below summarizes information about the electronic structure of the first eighteen elements. Complete the table by filling in all the missing information. You may use your periodic table.

		Electrons Present In:		
Element	Atomic #	n = 1	n = 2	n = 3
H	1	1	0	0
He	2	2	0	0
Li	3	2	1	
Be	4			
B	5			
C	6			
N	7			
O	8			
F	9			
Ne	10			
Na	11			
Mg	12			
Al				
Si				
P				
S				
Cl				
Ar				

*Intermediate Difficulty

Beyond the first eighteen elements, the rules governing electron placement become more complex. A more complicated model of electronic structure, one that considers orbitals and energy sublevels, is required to explain these more complex rules. These topics are beyond the scope of this book but are treated in nearly all introductory chemistry courses. Here, rules that can be used to predict most electron placements up to atomic number 38 are presented and practiced.

Beyond the first eighteen elements, electrons enter the electron shells in the following sequence: The next two electrons (#19 and #20) enter the fourth electron shell. The next ten electrons (#21–#30) enter the third electron shell and fill it to capacity. The next six electrons (#31–#36) enter the fourth electron shell. Finally, the next two electrons (#37 and #38) enter the fifth electron shell.

Questions

78. Look up the chemical symbol and atomic number for each element listed in the table below. Then work out the total number of electrons present in each atom and the number of electrons present in each electron shell.

				Electrons Present in Energy Level				
Element	**Symbol**	**Atomic Number**	**# Electrons**	**n = 1**	**n = 2**	**n = 3**	**n = 4**	**n = 5**
Potassium								
Calcium								
Iron								
Krypton								

79. In the space below, sketch an atom of potassium. Use a circle to represent each energy level.

*Intermediate Difficulty

As you can see from your periodic table, there are many elements beyond atomic number 38. The electronic structure of these elements is more complex and, therefore, they will not be considered further in this book. (Most of the chemistry that is necessary for other science areas, such as biology, involves the smaller elements.) For those elements with atomic numbers 1 to 38, the rules you have been given predict the electronic structure, with only minor exceptions. The rules are summarized in the table below.

Electrons	Electron Shell Entered
1–2	1
3–10	2
11–18	3
19–20	4
21–30	3
31–36	4
37–38	5

Questions

For each of the elements listed below, sketch an atom, showing how many protons and neutrons are present in the nucleus and how many electrons are present in each energy shell. Explain how you determined the structure for each atom.

80. $^{32}_{15}P$

81. $^{25}_{12}Mg$

82. $^{52}_{24}Cr$

The Atom: An Overview

A simplified model of the atom has been presented in Chapter 1. Try to create a mental picture of the atom before continuing on to further topics. This will help you organize and retain what you have learned. The material in the following paragraphs is intended to help you create such a picture in your mind. This is a challenge in the case of the atom, because the structures are so extremely tiny that atoms cannot be seen directly. Our model of the atom has been constructed to explain and to be compatible with many experimental results. A mental picture of the atom should be compatible with these results as well.

The absolute size of an atom is estimated to be about 0.1 to 0.5 nanometers. A nanometer is 10^{-9} meters. (This is one-billionth of a meter, the result of dividing a meter into tenths nine successive times.) The length of the dash in the brackets after this sentence is about 1 mm. { - } About ten million atoms could be lined up across this line in single file. The actual mass of a proton or neutron (about one atomic mass unit or amu) has been estimated as 1.673×10^{-24} grams. Multiplying this weight by the mass number would give the weight of a given atom in grams. Clearly, the objects we have been studying are so small that it requires considerable imagination to develop a sense of them at all.

Now try to imagine that you are traveling in a special spaceship that takes you into the ultramicroscopic world rather than into the vast reaches of outer space. As you approach the atom, a cloud or smear of negative charge would loom up first. Several layers of uniquely shaped clouds could be present. After passing through the electron clouds, you would encounter mostly empty space. It would be necessary to magnify to higher levels even to see the nucleus of the atom. The size of the nucleus is about 10^{-15} meter. The diameter of the atom, mostly due to the electron cloud, would be about a million times greater than this. Since the nucleus is so small, yet contains almost all the mass of the atom, it is extremely dense. The density of a nucleus has been estimated at 10^{12} grams per cubic centimeter, which is about one trillion times the density of a substance like water. So if you could see to a small enough level, you might get a glimpse of the atomic nucleus. This tiny, ultradense structure is even harder to visualize than the atom of which it is a part.

Another way to gain some perspective on the atom is to imagine that we have taken an atom with a low atomic number and magnified it repeatedly, while maintaining proper proportions, until it is the size of a football field. The nucleus would then be the size of a gumdrop in the center of the field. The electrons would be about the size of small insects, such as fruit flies. A few of these would be buzzing around at high speed somewhere in the vicinity, often in the region of the bleachers. All else would be empty space.

Finally, it is worth noting that all the matter we see is composed of atoms of the various elements. The objects in our world that seem so substantial to us (even our own bodies) are, in fact, composed of tiny invisible particles called atoms, containing mostly empty space. This is as true for lead weights as it is for air or a feather.

The diagrams on the next page are further examples of ways artists have attempted to represent atoms. All show aspects of the atom that have been verified experimentally. As the previous discussion has indicated, no diagram is truly realistic. Try to imagine an even better picture for yourself.

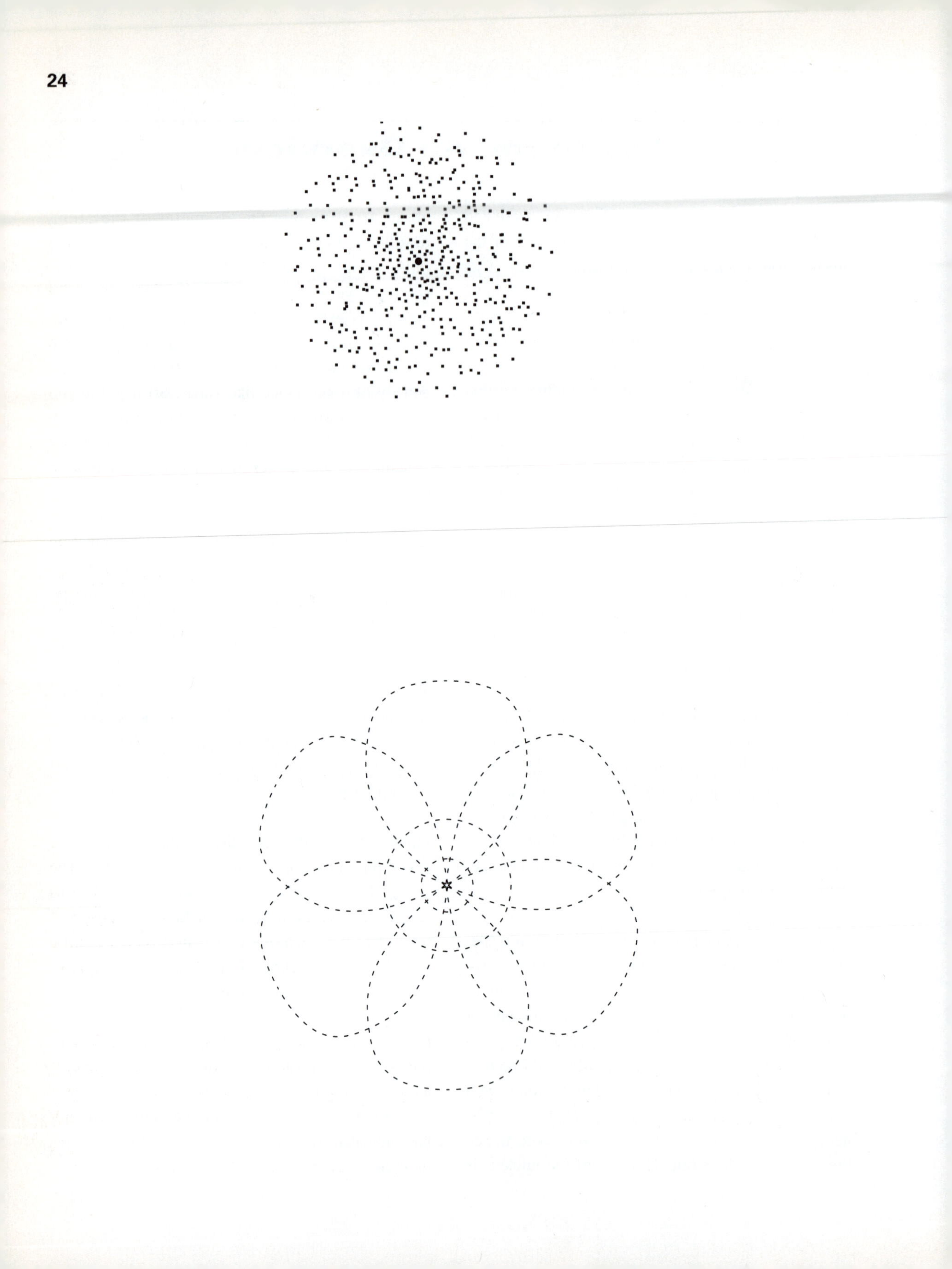

Answers

1. Yes.

Every person's body occupies a certain amount of space. Because body sh[illegible] volume of the body is most easily determined by water displacement. As sh[illegible] below, the rise in the level of the water when you enter a pool is your disp[illegible] weight of the body can be determined on a bathroom scale. Thus, the body [illegible]upies space, has mass, and is therefore composed of matter.

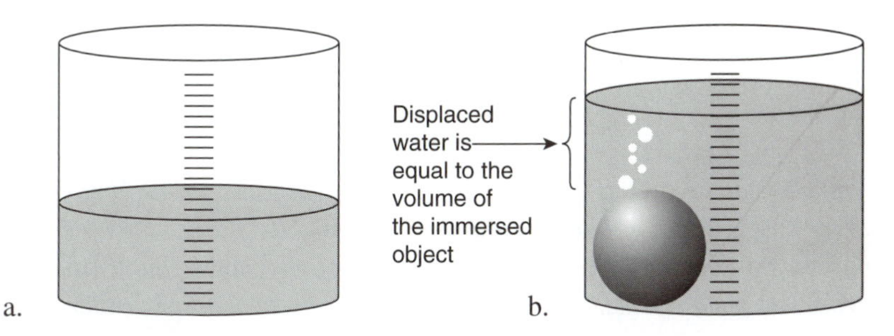

Volume measurement by water displacement. A measured amount of water is placed in a calibrated tank (a). The irregular object (such as the human body) is immersed in the water, causing the water level to rise (b). The increase in water level is measured to determine the volume of the immersed body.

2. Yes.

The chair takes up some room, and this is the space it occupies. Its actual volume could be measured by water displacement. You would weigh the chair to determine its mass. The chair occupies space, has mass, and is therefore composed of matter.

3. Yes.

Although the gaseous matter contained in the tank apparently was invisible, as are most gases, it did occupy the space of the tank. The fact that the tank weighed less after the vacuum pump was operated demonstrates that the tank was filled with something that had mass, probably some type of gas.

You could then use a higher-quality, more efficient vacuum pump for a longer time. If it succeeded in decreasing the weight of the tank still further, that would demonstrate that the tank had contained some residual matter after the first pump was used. Some of that residual matter was then removed by the second, more efficient pump.

, nitrogen, and neon are elements, because they do not decompose into simpler substances. on dioxide is not an element, because it decomposes into two components, carbon and oxygen. (Carbon and oxygen are themselves elements.) Therefore, carbon dioxide could not be one of the most simple building blocks and is not an element.

5. No.

Water is decomposed into two simpler substances during the course of this chemical reaction. Therefore, water is not an element.

6. No.

Sugar is not an element, because it decomposed into two other substances. Therefore, sugar could not be one of the simple building-block substances. Carbon dioxide and water were produced when sugar was burned. Information presented in previous questions showed that carbon dioxide and water are themselves not elements, because they can also be decomposed further.

7. Particle.

8. No.

The atom is defined as the smallest portion of an element that still retains the properties of the element. The smaller particles within an atom do not have the properties of the entire atom.

9. No.

Sugar is not an element (refer to question 6 if you have forgotten this). Because sugar is not an element, one particle of sugar cannot be an atom. An atom is defined as the smallest particle of an element that still has the properties of the element.

10. Atom. No.

Atoms are too tiny to see with the naked eye. Because 1 to 2 million atoms can fit across the diameter of a period, the size of one atom would be less than one-millionth of the diameter. This is far too small to see, even for a person with excellent vision.

11. The electron and the proton (a plus or minus sign indicates a charge).

12. The neutron. As its name implies, it is neutral and the charge is 0.

13. The proton and the neutron. (The weight of the electron is not significant.)

14. The electron. The electron's charge of –1 is opposite in sign to the proton's charge of +1. The charges are the same magnitude or size.

15. The electron. Its weight is about 1/2000 that of a proton; therefore, electrons make a negligible contribution to the weight of the atom.

16. Do not interact. The neutron is not charged and therefore neither repels nor attracts.

17. Attract because they have opposite charges.

18. Do not interact. A neutron has no charge and neither repels nor attracts.

19. Do not interact. Neither particle has a charge, and neither attracts or repels other particles.

20. Repel. Each electron carries a negative charge, and like charges repel each other.

21. Repel. Each proton carries a positive charge, and like charges repel each other.

22. True.

    All elementary particles are very small, as is the nucleus of the atom. Most of the atom consists of the empty space between the electron cloud and the nucleus of the atom.

23. The nucleus. It is positively charged because it contains the protons.

24. Nucleus.

25. An atom that contains three protons, three electrons, and four neutrons would be sketched as:

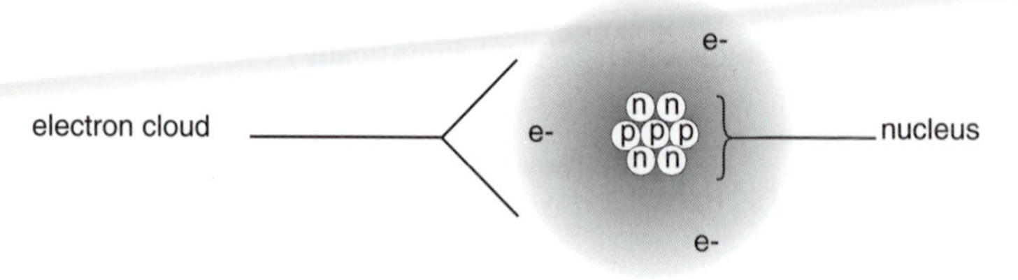

An Atom of Lithium

The number of elementary particles is represented correctly in a diagram of this type. The exact position of the particles is not known. The electrons are usually somewhere within the electron cloud around the nucleus. The size of the nucleus is much smaller than suggested in this diagram. If the nucleus were drawn to scale, it would be too tiny to see.

26. Six. The atomic number for carbon is 6. The atomic number is always equal to the number of protons in the atom.

27. In the nucleus.

28. No.

The atomic number establishes the identity of an element. Any nitrogen atom must have an atomic number of 7, and any atom with an atomic number of 7 must be nitrogen and could not be anything else. An atom with an atomic number of 8 is another element.

29. One.

30. Seven.

31. Protons.

32. Electrons.

33. Electrons.

The number of protons establishes the number of positive charges in the atom. Since electrons are negatively charged, the number of electrons must be the same as the number of protons to make the atom neutral.

34. Six protons; six electrons.

    This is determined by checking the atomic number. The atomic number gives the number of protons. Since the atom is neutral, the number of electrons must be the same.

35. Seven protons; seven electrons.

36. One electron.

    The atomic number of hydrogen is 1. This establishes that all hydrogen atoms have one proton. The number of electrons is equal to the number of protons, because the atom is neutral.

37. A rectangle on the periodic table contains the information about each element. The atomic number is located at the topmost position within the rectangle.

38. The name of the element is located just below the atomic number.

39. 20.

40. 11.

41. Iron has an atomic number of 26.

42. Magnesium has an atomic number of 12.

43. The symbols for the elements are:

Carbon	C
Hydrogen	H
Oxygen	O
Nitrogen	N

Note that the chemical symbol for each element is capitalized.

44. Phosphorus has an atomic number of 15. Its chemical symbol is P. Potassium has an atomic number of 19. Its chemical symbol is K.

The symbol for potassium is derived from its Latin name, kalium. Certain terms in clinical use refer to the Latin name of the element as well. For example, hypokalemia is a term that means the potassium concentration in the body is abnormally low.

45. The symbols for the elements are:

Sodium	Na
Sulfur	S
Iron	Fe

Note that when a chemical symbol consists of two letters, only the first letter is capitalized. The symbol for sodium is derived from its Latin name, natrium. The symbol for iron is derived from its Latin name, ferrum.

46. Chlorine. Cl is the symbol for this element. There are 17 protons and 17 electrons in a chlorine atom.

47. The symbols and atomic numbers are included in the table below.

Element	Symbol	Atomic #	Biological Roles
Hydrogen	H	1	Present in almost all organic molecules
Carbon	C	6	Forms the backbone of all organic molecules
Nitrogen	N	7	Found in all amino acids, proteins, and nucleic acids
Oxygen	O	8	Present in varying amounts in many organic molecules
			Essential gas for aerobic metabolism
Sodium	Na	11	Essential in tissue fluids and nerve impulse conduction
Magnesium	Mg	12	Required for many reactions in cells
Phosphorus	P	15	Found in all nucleic acids, some lipids, and adenosine tri-phosphate (ATP)
Sulfur	S	16	Present in some amino acids
Chlorine	Cl	17	Present in tissue fluids
Potassium	K	19	Essential in tissue fluids and nerve impulse conduction
Calcium	Ca	20	Involved in nerve impulse transmission, muscle contraction, and strengthens bone
Iron	Fe	26	Present in hemoglobin, which transports oxygen
Zinc	Zn	30	Required for many reactions in cells

48. The atomic number is 8. The mass number is 18.

49. 10 neutrons.

50. The mass number minus the atomic number is equal to the number of neutrons.

Mass number	=	# protons	+ # neutrons
– Atomic number	=	# protons	
	=		# neutrons

51.

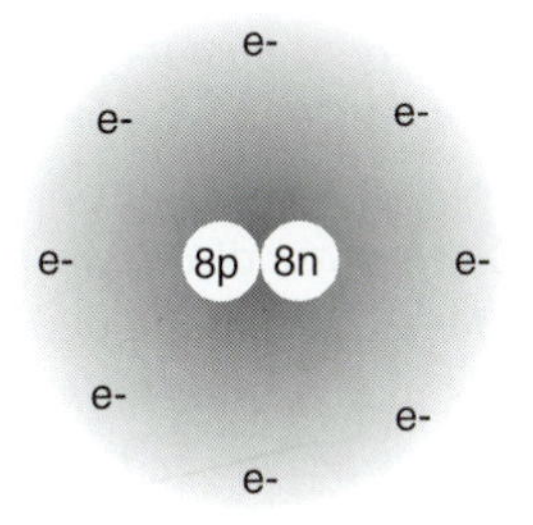

Both atoms have an atomic number of 8 and are therefore atoms of oxygen. Each has eight protons and eight electrons. The atoms have different numbers of neutrons. The atom shown here has eight neutrons, and the one in the example at the top of page 13 has ten neutrons.

52.

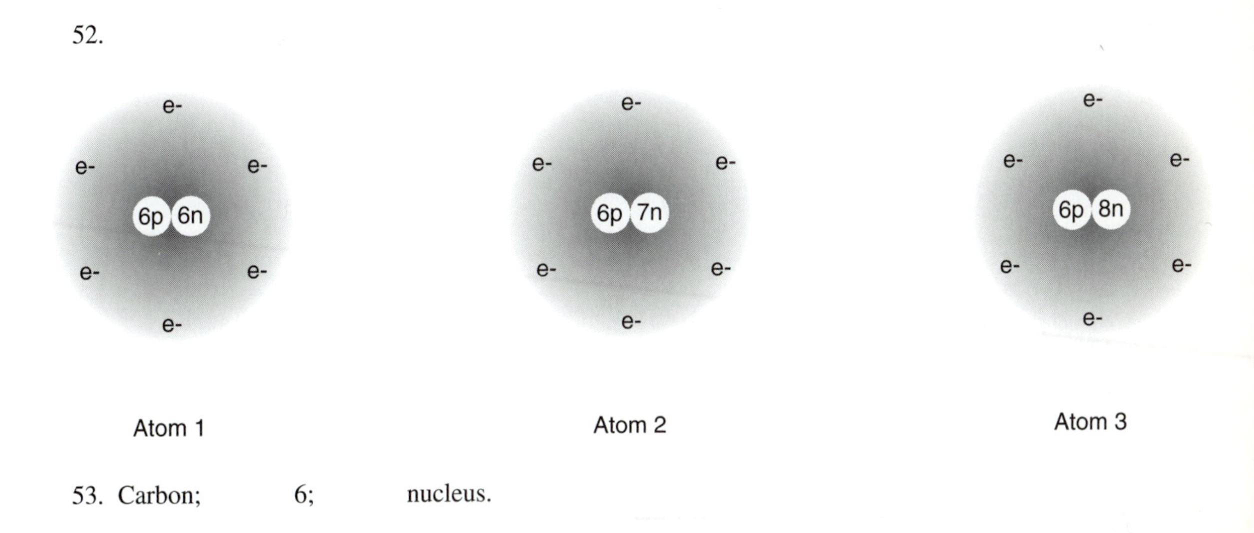

53. Carbon; 6; nucleus.

54. Carbon; 7; Carbon; 8.

55. Isotopes

Naturally occurring carbon is a mixture of these three isotopes. About 99% of all carbon atoms contain six neutrons, (like atom 1). Most of the remaining 1% of all carbon atoms contain seven neutrons (like atom 2). These two isotopes are stable. Atoms with eight neutrons (like atom 3) occur in trace amounts, much less than 1%. Carbon atoms with eight neutrons are unstable and undergo radioactive decay.

56. Same; same; different.

57. Unstable

Unstable isotopes are radioactive. During radioactive decay, the nucleus of the atom discharges particles and energy. High doses of radioactivity can be very toxic, because the high energies released during some radioactive decay processes can change and damage essential molecules in cells, including its genetic material. There are also many beneficial uses of radioisotopes. Some radioisotopes are used in medicine for diagnosis and treatment. Certain radioisotopes are also important in research, where they are used as tracers (a way of tagging certain molecules). During radioactive decay, the original atom may be changed to a different element. Such nuclear reactions will not be considered further in this book. In contrast, chemical reactions (covered extensively in future chapters) involve the electrons of atoms.

58. Isotopes

59.

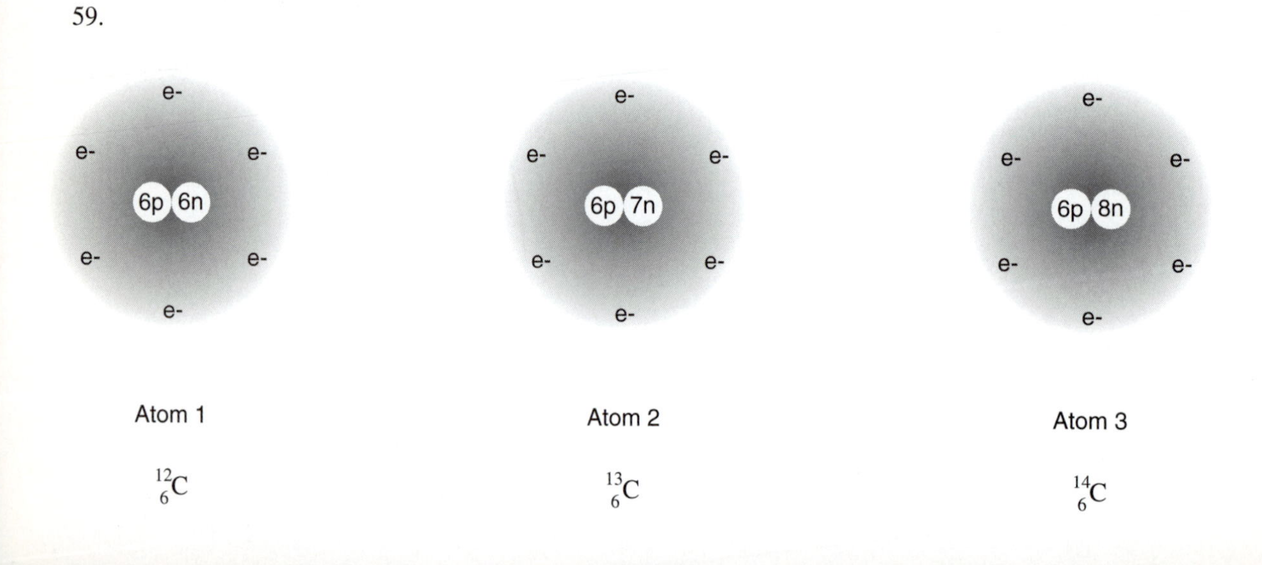

60. Isotopes; carbon 13; carbon 14.

61.

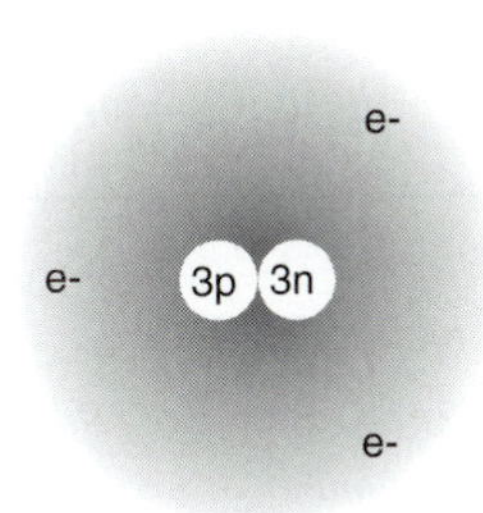

This isotope of lithium is called lithium 6, for its mass number, and accounts for 8% of the lithium atoms found in nature. The remaining 92% of lithium atoms contain four neutrons (and are called "lithium 7"). You sketched this more common lithium isotope for question 25 (on page 7). Both lithium isotopes are stable.

62. The atomic weights of the elements are:

Hydrogen	1.008
Carbon	12.01
Oxygen	16.00
Magnesium	24.31
Copper	63.54

63. Carbon, hydrogen, and oxygen.

64. Hydrogen, carbon, and oxygen are likely to have only one isotopic form that predominates in natural samples. These elements have atomic weights close to whole-number values. If there is only one common isotope, the atomic weight for the element will be close to the mass number for that most common isotope.

65.

Element	Atomic Number	Mass Number	# of Neutrons
Hydrogen	1	1	0
Carbon	6	12	6
Oxygen	8	16	8

The number of neutrons is calculated by subtracting the atomic number from the mass number. For example, the mass number of oxygen is 16. This gives the total number of neutrons plus protons. The atomic number of 8 is equal to the number of protons. Subtraction of the atomic number from the mass number gives the answer of 8. For each of these elements, 99% or more of the atoms are the common isotope described here.

66. Magnesium, with an atomic weight of 24.31, and copper, with an atomic weight of 63.54, have atomic weights that are not close to whole-number values. This is the result of having two or more isotopes that are relatively common. The atomic weight of the element is then between the mass numbers of the common isotopes.

67.

Isotope	# Protons	# Electrons	# Neutrons
$^{63}_{29}Cu$	29	29	34
$^{65}_{29}Cu$	29	29	36

The atomic number of 29 is equal to the number of protons. Since unreacted atoms are neutral, the number of electrons must be equal to the number of protons. The mass number (either 63 or 65, depending on the isotope) gives the total of neutrons plus protons. Subtraction of the atomic number from the mass number gives the number of neutrons.

68.*Copper with a mass number of 63 (copper 63).

We know it is the most prevalent form, because the atomic weight of 63.54 is closer to 63 than it is to 65. This is the expected result if natural samples of copper contain more atoms of copper 63. (Measurements of the amount of each isotope in copper samples show that copper 63 accounts for about 69% of the sample, and copper 65 makes up about 31%.)

69. No.

***The atomic weight of 63.54 amu for copper results from the fact that a sample of copper is a mixture of two isotopic forms, each with a different weight. About 69% of the atoms have a mass number of 63, and the rest have a mass number of 65. The atomic weight is between these values, because both types of atoms are mixed together in natural samples. Therefore, the atomic weight is not the actual weight of any one atom but the average weight of naturally occurring copper atoms.*

70. 2; Yes; No.

For the first electron shell, $n = 1$. Inserting this value into the formula, $2n^2$, gives the answer, $2 \times 1^2 = 2 \times (1 \times 1) = 2 \times 1 = 2$.

The first shell could not contain four electrons, because its maximum capacity is two. The shell can contain fewer electrons than the maximum but cannot contain more.

71. 8, calculated as $(2 \times 2^2) = 8$. Yes; No.

72. 18, calculated as $(2 \times 3^2 = 18)$; No. Twenty-six exceeds the maximum capacity of eighteen for this electron shell.

73. Carbon
Atomic # = 6
\# electrons = 6

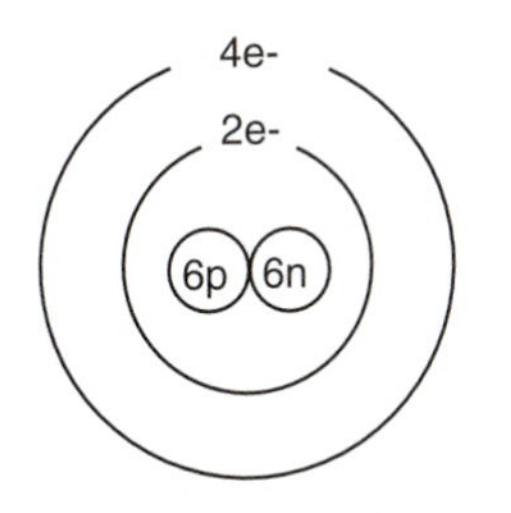

74. Chlorine
Atomic # = 17
\# electrons = 17

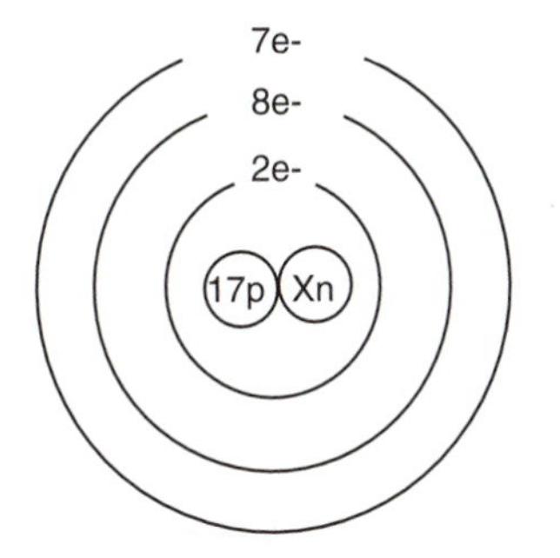

75. Hydrogen
Atomic # = 1
electrons = 1

76. Sodium
Atomic # = 11
electrons = 11

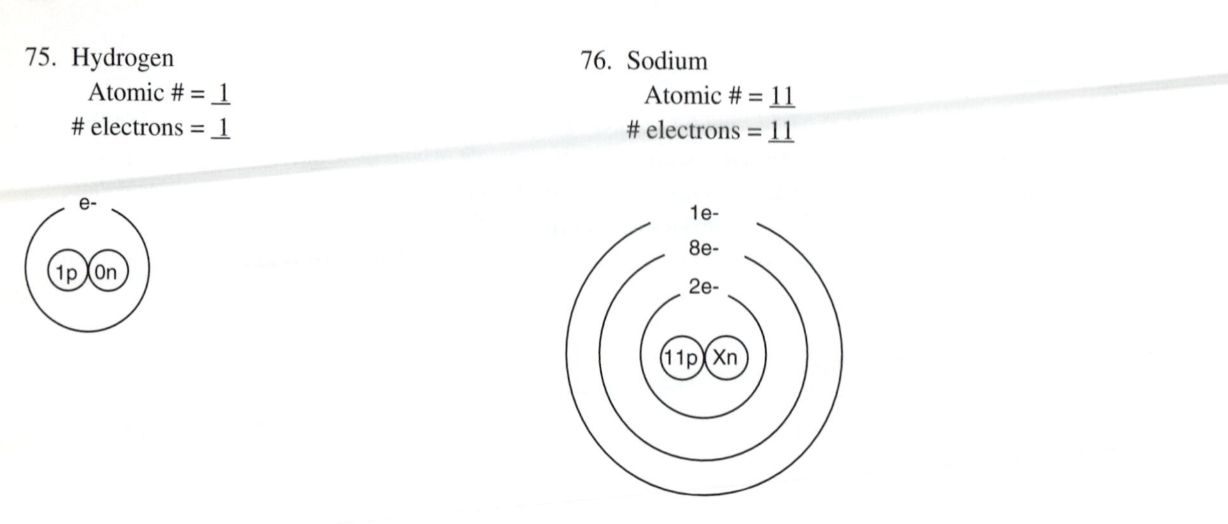

77.

Element	Atomic #	Electrons Present In:		
		n = 1	n = 2	n = 3
H	1	1	0	0
He	2	2	0	0
Li	3	2	1	0
Be	4	2	2	0
B	5	2	3	0
C	6	2	4	0
N	7	2	5	0
O	8	2	6	0
F	9	2	7	0
Ne	10	2	8	0
Na	11	2	8	1
Mg	12	2	8	2
Al	13	2	8	3
Si	14	2	8	4
P	15	2	8	5
S	16	2	8	6
Cl	17	2	8	7
Ar	18	2	8	8

78.

Element	Symbol	Atomic Number	# Electrons	Electrons Present in Energy Level				
				n = 1	n = 2	n = 3	n = 4	n = 5
Potassium	K	19	19	2	8	8	1	0
Calcium	Ca	20	20	2	8	8	2	0
Iron	Fe	26	26	2	8	14	2	0
Krypton	Kr	36	36	2	8	18	8	0

Potassium and calcium are the first elements (beyond the 18 considered earlier) that follow more complex rules for electron placement. The nineteenth and twentieth electrons enter the fourth energy shell, even though the third energy shell is not yet filled. After there are two electrons in the fourth shell, any additional electrons enter the third energy shell, until it is filled with eighteen. In the case of iron, six additional electrons (beyond the eight already present) have entered the third shell for a total of fourteen. The element krypton has a total of thirty-six electrons. After the third shell has been filled, there are still six electrons to be assigned. These enter the fourth shell, joining the two already placed there.

79.

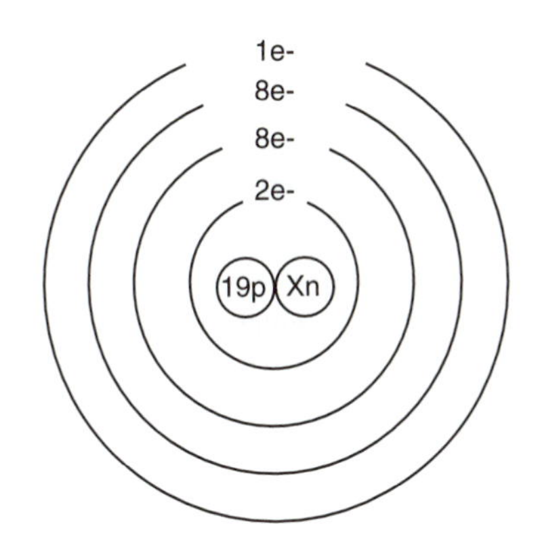

80. $^{32}_{15}P$ 81. $^{25}_{12}Mg$ 82. $^{52}_{24}Cr$

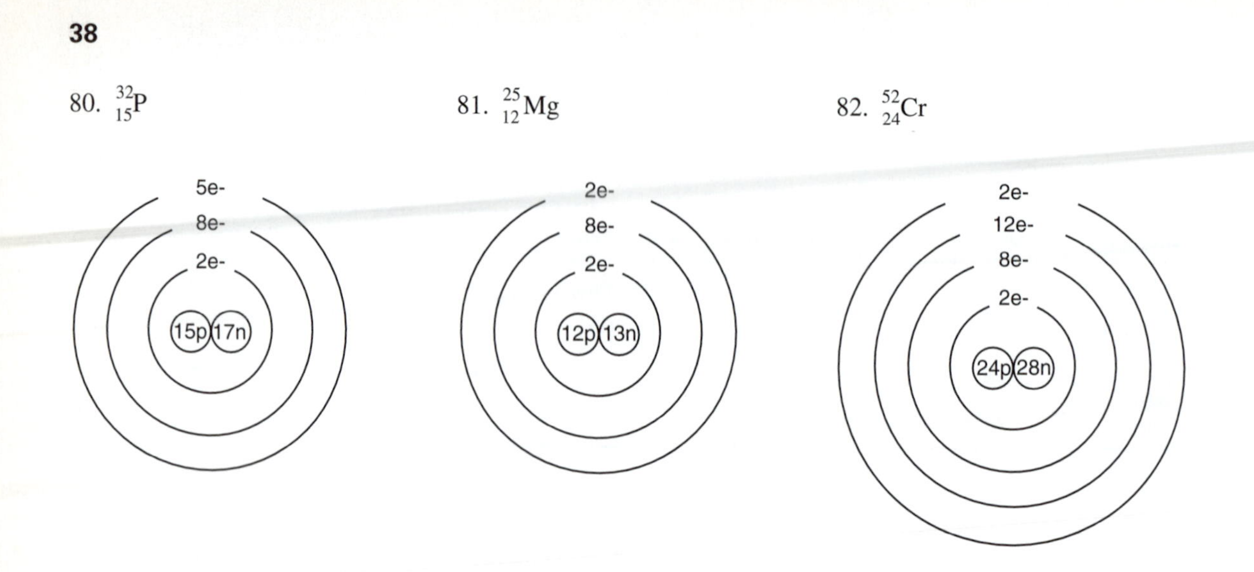

80. Phosphorus has an atomic number of 15. This tells us that phosphorus has 15 protons and 15 electrons. The first two electrons fill the first shell. The next eight electrons fill the second shell. The remaining five electrons are located in the third shell. Subtracting the atomic number of 15 from the mass number of 32 shows that this isotope contains 17 neutrons.

81. Magnesium has an atomic number of 12; thus, it has 12 protons and 12 electrons. The first two electrons fill the first shell, the next eight electrons fill the second shell, and the remaining two electrons occupy the third shell. There are 13 neutrons in this isotope, obtained by subtracting the atomic number of 12 from the mass number of 25.

82. Chromium has an atomic number of 24; thus, it has 24 protons and 24 electrons. The first two electrons fill the first shell, the next eight electrons fill the second shell, and the next eight electrons enter the third shell. The next two electrons enter the fourth shell, which begins to fill before the third shell is completely filled. The remaining four electrons enter the third shell, which then contains twelve electrons and is still not filled to capacity. Subtracting the atomic number of 24 from the mass number of 52 gives 28 for the number of neutrons in this isotope.

Chapter Test: The Atom

The questions in this chapter test evaluate your mastery of all the objectives for this unit. Take this test without looking up material in the chapter. Questions that test intermediate objectives are indicated by an asterisk, and you may omit them if you did not cover those topics in the chapter. After you have completed the test, you can check your work by using the answer key located at the end of the test.

1. An irregular object that weighs 1 kilogram is composed of

   a. energy.

   b. complex substances.

   c. matter.

   Explain why you chose this answer. ______________________________

   ____

2. Which elementary particles have a charge of –1 and a mass that is almost zero? ____________

3. If you know the atomic number of an element, you also know the number of ________ and ________ in unreacted (neutral) atoms of that element.

4. How do a proton and an electron interact? ______________________________

5. These questions pertain to an atom that can be represented as ${}^{7}_{3}Li$. This atom has ____ protons, ____ electrons, and ____ neutrons. In the space below, sketch an atom of this type. Show how many electrons are present in each electron shell.

6. Elements with the same atomic number but different mass numbers are ________.

7. Radioisotopes are

   a. stable.

   b. unstable.

8. State the formula used to calculate the maximum number of electrons in an electron shell, and explain what the formula means. ______________________________

9. Calculate the maximum number of electrons that can occupy the third electron shell. Show your work.

____

10. Look up the chemical symbol and atomic number for each element listed in the table below. Then work out the total number of electrons present in each atom and the number of electrons present in each electron shell.

				Electrons Present in Energy Level				
Element	**Symbol**	**Atomic Number**	**# Electrons**	**n = 1**	**n = 2**	**n = 3**	**n = 4**	**n = 5**
Carbon								
Nitrogen								
Chlorine								
Iron*								

11–17. For the next series of questions, indicate whether the statement about atoms is true or false. If the statement is false, explain why.

11. Atoms are mostly empty space.

True ______ False ______

____

12. Atoms can be seen with the aid of a magnifying glass.

True ______ False ______

____

13. The electron cloud is negatively charged.

True ______ False ______

____

14. Most of the mass is in the nucleus.

True ______ False ______

____

15. Electrons have definite locations and orbits.

True ______ False ______

____

16. The nucleus is electrically neutral.

True ______ False ______

____

17. Atomic nuclei are not very dense.

True ______ False ______

____

18–20. For the next series of questions, define and explain the term as completely as you can.

18. Element: ____

19. Atom: ____

20. Periodic table of the elements: ____

____

21.*The atomic weight for chlorine is 35.45. Explain why the atomic weight for this element is not close to a whole-number value.

____

____

22.*In the space below, sketch an atom of bromine with an atomic number of 35 and a mass number of 80.

Show the number of electrons in each electron shell.

Answers

1. c. matter.

   The weight of one kilogram indicates that the object has mass. Its irregular shape implies that it occupies space. Matter is anything that occupies space and has mass.

2. electrons

3. electrons and protons Note: answers may be in either order.

4. They attract each other.

5. 3 protons, 3 electrons, and 4 neutrons.

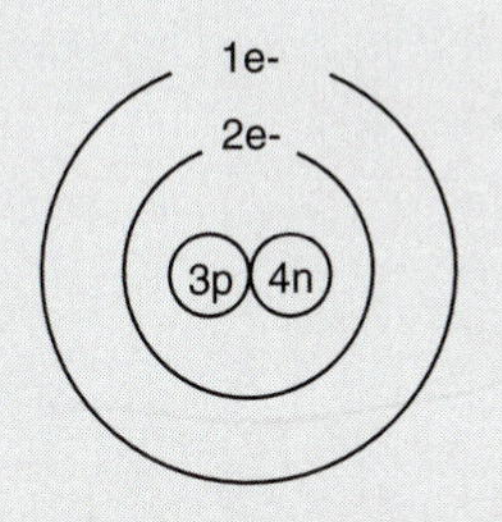

6. isotopes

7. b. unstable.

8. $2n^2$ = maximum number of electrons in the shell, where n = the principal energy level

9. When n = 3, the calculation using the formula stated in the last question is:
   $2(3^2) = 2 \times 3 \times 3 = 18$ electrons

10.

				Electrons Present in Energy Level				
Element	**Symbol**	**Atomic Number**	**# Electrons**	**n = 1**	**n = 2**	**n = 3**	**n = 4**	**n = 5**
Carbon	C	6	6	2	4	0	0	0
Nitrogen	N	7	7	2	5	0	0	0
Chlorine	Cl	17	17	2	8	7	0	0
Iron*	Fe	26	26	2	8	14	2	0

11. True

12. False

Atoms are estimated to have such a small size that a million would fit across a printed period. Such small structures are not visible even if magnified by a hand lens. In fact, they cannot be observed even with the best light microscopes.

13. True

14. True

15. False

The electrons are best imagined either as moving very rapidly or as a smear of negative charge described as the electron cloud. The orbits are not locations that the electrons must traverse but rather probable locations of the electrons.

16. False

The nucleus contains both protons and neutrons. The neutrons are not charged. The protons carry a positive charge; thus, the nucleus containing them is also positively charged.

17. False

    Atomic nuclei contain both protons and neutrons, the elementary particles with significant mass. The nucleus is extremely small. Since nearly all the mass of the atom is contained in a tiny structure, it is also very dense.

18. Elements are substances that cannot be converted to less complex substances by chemical reactions. The elements are the basic building blocks of more complex forms of matter.

19. An atom is the smallest particle of an element that still retains the properties of that element.

20. The periodic table of the elements is an arrangement of the elements according to their atomic numbers. A great deal of information about the elements is presented in the table.

21.* An element that has an atomic weight between whole numbers typically contains a mixture of two or more different isotopes. Atoms of two or more different weights are mixed together in samples of the element. That is probably the case for chlorine. (In fact, chlorine *is* a mixture of isotopes. About 75% of the atoms are an isotope with a mass number of 35. The other 25% have a mass number of 37. The weighted average of these values is 35.45.)

22.*

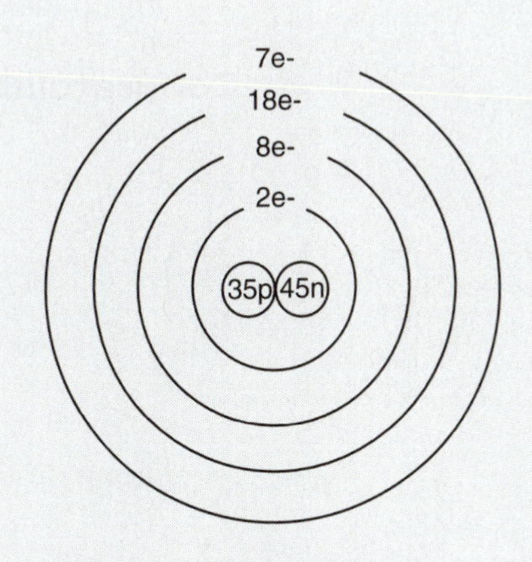

Self-Evaluation

You have now taken the chapter test on the atom and checked your answers. If you are like many students, you now want to know what your grade on this test would be. An adjustment of attitude may be necessary, because there are no grades on this test. The goal of this textbook is to help you master the basic subject matter of chemistry. If you have mastered the subject matter of this chapter, your score on the chapter test should be 100%. But don't be discouraged if you missed a few questions. Treat the chapter test as a learning experience that can help you identify areas you need to study further.

Suppose you answered question 10 incorrectly. This question is a complex one involving several of the objectives stated at the beginning of the chapter. If you missed some of the chemical symbols, further review and drill on those symbols would be appropriate until you have memorized them. If you worked out the electronic structure of the atoms incorrectly, review the sections on the electronic structure of atoms and rework the questions on that topic in the chapter itself. When you have reviewed and restudied the topics that gave you problems, retake the chapter test. Your score should improve.

When you have mastered the material in this chapter, you are ready to move to the next topic. Chemistry is a cumulative subject. Since later topics build on your understanding of the material presented previously, careful work now will ensure your continued success. There is one exception to this. The intermediate and advanced material can be omitted, since future topics do not depend on it. Individualize your chemistry study to include the topics essential for your own academic program.

When you can answer all the questions within this chapter and in the chapter test at the end, you are ready to move on.

Congratulations!

Chapter 2

Ionic Bonding

Objectives

1. Define the following terms and apply the definitions correctly: compound, formula, molecule, valence shell, valence electrons, ion, anion, cation, ionization, metal, nonmetal, ionic bond, oxidation, and reduction.
2. Write and interpret chemical formulas that represent the composition of compounds.
3. State and explain the octet rule. Use the octet rule to predict what elements will be stable and what elements will be reactive.
4. Use the octet rule to predict what elements will ionize and what elements will not. Predict the net charge of ions that do form.
5. Predict which combinations of ions can form bonds.
6. Use the position of an element on the periodic table to predict the number of electrons in the valence shell.
7. *Predict the ratios of positive and negative ions in ionically bonded compounds.

Compounds, Formulas, and Chemical Notation

Compounds are substances composed of two or more elements combined in a definite proportion by mass. Thus, compounds are substances that can be decomposed into simpler substances by chemical reaction. A compound can be decomposed into the elements that are present within it. The atoms of the elements form chemical bonds with each other within the compound. Two or more atoms linked together by chemical bonds form a molecule.*

Questions

1. Water can be decomposed by an electrical current, a process called electrolysis. Hydrogen gas and oxygen gas are produced in the decomposition process. Hydrogen and oxygen are produced in a 2:1 ratio. Based on this information, is water a compound? Explain.

   Yes ______ No ______

2. Atoms interact with each other within compounds by means of ____________ and join to form ____________.

3. A substance is decomposed and produces equal amounts of hydrogen and oxygen. Could this substance be water? Explain.

   Yes ______ No ______

4. You prepare a saltwater solution by dissolving some table salt in pure water. Is the saltwater solution a compound? Explain.

   Yes ______ No ______

*The term "molecule" is used to describe atoms linked together by covalent bonds, a type of bonding that involves sharing electrons (covered in Chapter 3). The chemical formula therefore represents the atoms within a molecule, as well as the proportion of elements within the compound. Ionic bonding (covered in this chapter) results from electron transfer. Individual molecules do not form, and the formulas simply represent the proportions of the elements that are present in the ionically bonded compound.

Every compound can be represented by a formula that tells what elements are present in the compound and the proportions of each. For example, $NaHCO_3$ is the formula for sodium bicarbonate. The subscript 3 after the symbol for oxygen means that each sodium bicarbonate molecule contains three atoms of oxygen for every one atom of sodium, hydrogen, and carbon. (When no subscript follows a chemical symbol, a 1 is understood.)

Questions

5. Sodium bicarbonate (also called baking soda) contains __________ different elements in the proportions of: __________ sodium to __________ hydrogen to __________ carbon to __________ oxygens.

6. How many atoms are present in each sodium bicarbonate molecule, and how are the atoms linked?

7. State whether each of the formulas in the table below represents an element or a compound. Explain your reasoning.

Formula	Element or Compound
H	
CO	
H_2O	
O	
$C_6H_{12}O_6$	

Stability and the Octet Rule

The elements listed in the table below include some of the noble gases, previously known as the inert gases. These elements are very stable and tend to remain in the elemental state. They have very little tendency to undergo chemical reactions to form compounds. Careful study of this group provides clues about the electronic structures that correlate with chemical stability.

		Number of Electrons in Electron Shell				
Element	Atomic Number	n = 1	n = 2	n = 3	n = 4	n = 5
He	2	2	0	0	0	0
Ne	10	2	8	0	0	0
Ar	18	2	8	8	0	0
Kr	36	2	8	18	8	0
Xe	54	2	8	18	18	8

Questions

8. In helium, electrons are located in the ____________ electron shell. This electron shell contains ____________ electrons. The maximum capacity of this energy level is ____________ electrons. Is this electron shell full?

   Yes ______ No ______

9. Which electron shells (principal energy levels) are occupied in atoms of the element neon? ____________ The maximum capacity of the outer electron shell (n = 2) is ____________ electrons. This electron shell contains ____________ electrons. Could this energy level accept any more electrons?

   Yes ______ No ______

10. Which electron shells (principal energy levels) are occupied in atoms of the element argon? ____________ Which are occupied in atoms of krypton? ____________ Which are occupied in atoms of xenon? ____________

11. The outermost shell occupied in argon is principal energy level ______. It contains ______ electrons. The outermost shell occupied in krypton is ______. It contains ______ electrons. The outermost shell occupied in xenon is ______. It contains ______ electrons.

The outermost electron shell that contains electrons in an atom is called the valence shell. Electrons that occupy the valence shell are called valence electrons. Elements that contain eight electrons in the valence shell tend to be stable and unreactive. Elements that do not have eight valence electrons tend to be less stable and more reactive. This is a statement of the octet rule. The octet rule can be used to predict and explain the chemical behavior of many elements. Note that in the case of helium, stability is achieved with just two electrons (the first electron shell has a maximum capacity of 2).

Questions

12. Determine the number of valence electrons for each of the atoms listed in the table below. Predict whether atoms of each element would be stable or unstable. Explain.

Element	Symbol	Valence Shell	# Valence Electrons	Stable or Unstable
Helium				
Neon				
Argon				
Krypton				
Oxygen				
Hydrogen				
Magnesium				
Nitrogen				
Carbon				

13. Elements tend to be stable and unreactive when the valence shell contains ______ electrons. (The table in the previous question should help you answer this.)

14. The element radon, atomic number 86, has eight electrons in principal energy level 6. This is the outermost electron shell that contains electrons in this element. What is the valence shell for radon? ______ Would you expect this element to be chemically reactive? Explain your answer.

    Yes ______ No ______

15. When the valence shell does not contain ______ electrons, elements tend to be unstable and reactive.

Formation of Positive Ions

The table below is a partial listing of the elements from group IA. These elements are unstable and undergo vigorous chemical reactions. With the exception of hydrogen, these elements show very similar chemical behavior.

Element	Atomic Number	Valence Shell	# Electrons in Valence Shell
H	1	1	1
Li	3	2	1
Na	11	3	1
K	19	4	1
Rb	37	5	1

Questions

16. The elements in group IA all have ______ valence electron(s). Is this a stable configuration? Explain your answer.

    Yes ______ No ______

17. Use the periodic table to look up the elements listed above. Where are they located and how are they arranged?

18. Look up the element cesium in the periodic table. Where is it located? ______

19. Cesium would have ______ valence electron(s). Would you predict that Cs would be a stable element like the noble gases? Explain your prediction.

    Yes ______ No ______

20. Elements with eight valence electrons tend to be ____________, whereas elements with one valence electron tend to be ____________.

During chemical reactions, elements from group IA generally lose one electron, thus becoming stable. This reaction empties the valence shell and exposes an octet in the electron shell below. After loss of the valence electron, the atom no longer has equal numbers of protons and electrons, so it is now charged. Atoms that have developed a net charge due to the loss or gain of electrons are called ions. An example of this type of reaction is shown below.

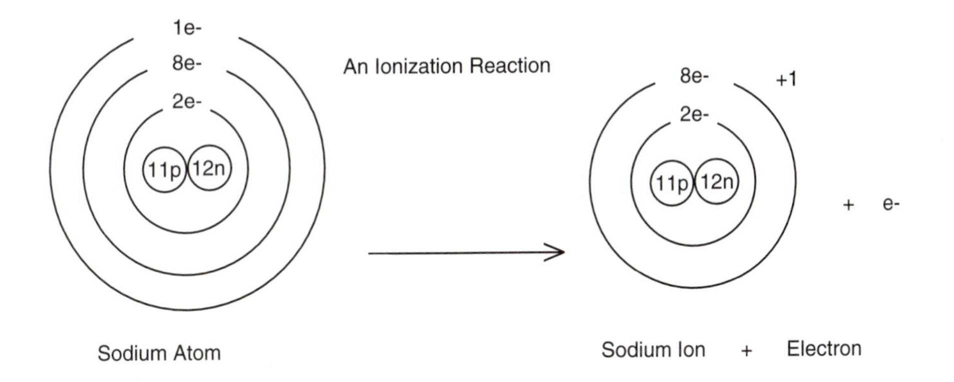

Questions

21. The atomic number of Na is ______, and sodium atoms have ______ protons and ______ electrons. The overall charge on the Na atom is therefore ______, and the atom is neutral.

22. In the diagram above, the chemical reaction is represented by an arrow. During this chemical reaction, an ____________ is lost by the Na atom. This is an example of an ____________ reaction.

23. After ionization of the Na atom, there are ______ protons and ______ electrons. The net charge on this structure is now ______. An atom that has developed a net charge due to the loss or gain of electrons is called an ____________.

24. In the space below, diagram a Li atom undergoing an ionization reaction of the same type as that shown for Na. Label the diagram.

25. The product of lithium ionization is a ____________ with a net charge of ____________.

26. The product of potassium ionization would be a ____________ with a net charge of ____________.

The table below is a partial listing of the elements from group IIA. These elements are reactive and exhibit similar chemical behavior. During chemical reactions, elements from group IIA generally lose two electrons (from the valence shell), thus becoming stable. As in the previous examples, after loss of the valence electrons, the atom no longer has equal numbers of protons and electrons. It is then charged, and forms positive ions.

Element	Atomic Number	Valence Shell	# Electrons in Valence Shell
Be	4	2	2
Mg	12	3	2
Ca	20	4	2
Sr	38	5	2

Questions

27. All the elements listed above have ______ electrons in the valence shell. Is this a stable configuration? Explain your answer.

    Yes ______ No ______

28. Use the periodic table to look up the four elements listed above. Where are they located and how are they arranged?

29. In the space below, diagram an ionization reaction for an atom of the element Mg and label it.

30. The atomic number of Mg is ______. Magnesium atoms have ______ protons and ______ electrons. The overall charge on the Mg atom is therefore ______ and the atom is neutral.

31. After ionization of the Mg atom, there are ______ protons and ______ electrons. The net charge on this ion is now ______.

Elements with three valence electrons tend to lose those electrons in chemical reactions. They form positive ions with a net charge of positive three. Positive ions are also called cations.*

Questions

32. The element aluminum has an atomic number of ______ and is represented by the symbol ______ . The valence shell for aluminum is ______.

33. In the space below, sketch the ionization reaction for Al.

34. Al atoms that have a mass number of 27 have ______ electrons, ______ protons, and ____________ neutrons. Each electron has a charge of ____________, each proton has a charge of ____________, and each neutron has a charge of ____________.

35. After ionization, the Al would have ______ electrons, ______ protons, and ______ neutrons. The net charge on the Al ion would be ____________. Since the aluminum ion has a positive charge, it could also be called a ____________. If subjected to an electrical current, it would move toward the ____________ terminal or ____________, since opposite electrical charges attract each other.

*The origin of the term "cation" is as follows. Positive ions move toward the cathode (the negative terminal) during electrolysis. Similarly, negative ions are called anions, because they move to the anode, (the positive terminal) during electrolysis. Electrolysis reactions use an electrical current to provide energy to a chemical reaction. The details of how this occurs and the equipment that is used are not covered in this text and need not concern you at this time. Since the terms are still in use, you should know them. A cation is a positive ion, and an anion is a negative ion.

As a general rule, elements that have one, two, or three valence electrons can form positive ions (cations). The valence electrons are lost, exposing an octet of electrons in the shell below. The cations formed in this way are quite stable.

Questions

36. Complete the table below. For each atom, indicate which principal energy level is the valence electron shell. State how many electrons the valence shell contains. Then work out the structure of the ion that could form from each of these atoms during a chemical reaction. Which shell is the valence shell in the ion? How many electrons does it contain?

	Atom		Ion	
Element	Valence Shell	# Valence Electrons	Valence Shell	# Valence Electrons
Li	2	1	1	2
Na	3	1	2	8
Mg	3	2	2	8
Al	3	3		
K	4			
Ca				
Rb				
Sr				

37. Prior to reacting, the atoms considered here had ______, ______, or ______ valence electrons. With the exception of lithium, after ionization, the valence shell contains ______ electrons.

38. Recall the octet rule, and explain how it can be used to predict the results described in the previous questions.

Formation of Negative Ions

Elements with six or seven valence electrons are not stable. Such elements are reactive, and during chemical reactions, they accept electrons until an octet is formed in the valence shell. The elements in the table below are examples of this group.

Element	Atomic Number	Valence Shell	# Electrons in Valence Shell
F	9	2	7
Cl	17	3	7
Br	35	4	7
I	53	5	7

Questions

39. Use the periodic table to look up the elements listed above. How are they arranged and to what group do they belong?

40. Cl atoms have ______ valence electrons and are unstable. If such an atom ______ one electron, it will then have eight electrons and will be ____________.

41. In the space below, sketch the reaction described in the previous question and label it. This reaction is a type of ____________.

42. After ionization, the chloride ion has ______ protons and ______ electrons. It now has a net charge of ______. Explain how you determined the net charge.

Elements with six or seven valence electrons complete their octet by accepting additional electrons. After accepting electrons, they form negative ions. Recall that negative ions are also called anions. Ions with full valence octets are stable.

Questions

43. In the space below, sketch a diagram showing the ionization process that converts fluorine atoms to ____________ ions.

44. The net charge on the fluoride ion is ____________. The net charge on the chloride ion is ____________. The net charge on the bromide ion would be ____________.

45. As a general rule, elements with seven valence electrons react by gaining ____________ electron(s) to create an octet in the valence shell. This produces ____________ with a net charge of ____________.

46. The elements oxygen, sulfur, and selenium all have six valence electrons. Look these elements up in the periodic table of the elements. How are these elements arranged, and where are they located?

47. How can elements of this group acquire an octet of electrons in the valence shell?

48. In the space below, diagram an ionization reaction that creates a valence octet for an atom of oxygen.

49. An oxygen atom has ______ protons, ______ electrons, and a net charge of ______. An oxide ion has ______ protons, ______ electrons, and a net charge of ______.

50. In general, elements with six valence electrons tend to ______ two electrons to create an octet of electrons in the valence shell. After such an ______ reaction, the net charge is ______ on the ion.

Oxidation and Reduction

Oxidation is defined as the loss of one or more electrons. Reduction is defined as the gain of one or more electrons. These definitions can be applied to the two types of ionization we have already discussed. Oxidation and reduction occur together during chemical reactions. The electrons lost by the element that is oxidized are gained by the element that is reduced. Oxidation/reduction reactions (also called redox reactions) are important in the biochemistry of all cells.

Questions

51. Elements with one, two, or three valence electrons tend to ____________ their valence electrons and form ____________. This type of ionization is referred to as __________________.

52. Elements with one, two, or three valence electrons tend to be listed at what part of the periodic table?

53. Elements with one, two, or three valence electrons generally function as electron ____________ in chemical reactions, and they become ____________ in the process of losing their valence electrons.

54. Elements with six or seven valence electrons typically ______ electrons to create a valence shell with an ____________ of electrons. This type of ionization is an example of a ____________ reaction.

55. Elements that have six or seven electrons function as electron ____________ and become __________________ as they accept electrons.

56. Elements that react by undergoing oxidation tend to be listed in what part of the periodic table?

57. Elements that react by becoming reduced tend to be listed in what part of the periodic table?

58. A redox reaction can be thought of as an electron transfer reaction, wherein an element that can donate electrons becomes ____________ and an element that accepts electrons becomes ____________.

As we have seen previously, the elements are arranged in patterns in the periodic table. Elements that have the same number of valence electrons are found in a column. The atomic number increases as we move down a column.

Questions

59. The elements with seven valence electrons are found in a ____________ labeled group ______ in the ____________ of the periodic table.

60. The elements with one valence electron are found in a ____________ labeled group ______ in the ______ of the periodic table.

61. State a general rule for using the periodic table when determining the number of valence electrons in atoms of a particular element.

62. In the chemical reactions we have examined so far, which elementary particles participated in chemical reactions?

63. Review the reaction of Na ionization. How many electron shells are occupied in the Na atom? ______ Which electrons are involved in the ionization process? ____________ Is this ionization an oxidation or a reduction? ____________

64. Review the reaction that forms Cl ions from Cl atoms. How many electron shells are occupied in Cl atoms? ______ Which electrons are involved in the ionization reaction? ____________ Is this ionization an oxidation or a reduction? ____________

65. Based on the examples above, which electrons participate in chemical reactions?

66. In general, elements exhibit similar chemical behavior if they have the same number of ____________. The number of valence electrons can be determined easily by ____________.

67. Explain how the octet rule can be used to predict which elements will ionize.

Elements with one to three valence electrons function as electron donors and become oxidized during chemical reactions. These elements are found in the left-hand portion of the periodic table and are usually metals. Elements with six or seven valence electrons function as electron acceptors and become reduced during chemical reactions. They are found in the right-hand portion of the periodic table and are usually nonmetals. Elements with four or five valence electrons (and an atomic number below 18) react so as to share electrons (a topic covered in the next chapter) and are usually nonmetals. Elements with eight valence electrons are unreactive nonmetals.

Questions

68. Complete the table below and explain the reasons for your answer.

Element	# Valence Electrons	Type of Reaction
C	4	Shares electrons
Na		
Ca		
O		
N		
Mg		

69. Mg tends to ____________ electrons during chemical reactions. After such a reaction, the atom has formed a ____________ with a net charge of ______. This electron transfer reaction is an example of ____________. Magnesium is a __________________ that is located __________________ in the periodic table.

70. Oxygen tends to ____________ electrons during chemical reactions. After such a reaction, the atom has formed an ____________ with a net charge of ____________. This electron transfer reaction is an example of __________________. Oxygen is a ____________ that is located __________________ in the periodic table.

Ionic Bonding

Opposite charges attract each other. Positive and negative ions interact in this way. The attractive force between ions of opposite sign is responsible for ionic bonding.

Questions

71. For each element listed below, determine whether atoms of that element can ionize. If ions can form, determine what the net charge would be by using the octet rule.

Element	# Valence Electrons	Ionization	Stable Ionic Form
Na	1	Yes	Na^+
C			
K			
Cl			
H			
Mg			
O			
Al			
N			

Use the information in the table above to answer the following questions.

72. Would ionic bonds form between Na and K ions? Explain.

Yes ______ No ______

73. Would ionic bonds form between Na and Cl ions? Explain.

Yes ______ No ______

74. Would ionic bonds form between Na and C? Explain.

Yes ______ No ______

75. Ions of the opposite sign (one positive and one negative) will be ______ to each other and could form an ______ bond. Ions of the same sign will ______ each other and ______ form ionic bonds.

A compound formed due to ionic bonding is neutral; yet it contains both positive and negative ions. The number of positive charges and the number of negative charges within the compound therefore must be equal. This determines the proportions of the elements present in a compound and is represented in the formula for the compound.

Questions

Refer to the table of ions in question 71 as you answer the following questions.

76. What is the charge on the Na ion? ______ What is the charge on the Cl ion? ______ Can these two ions bond to form a compound? Explain.

    Yes ______ No ______

77. If one Na ion and one Cl ion form an ionic bond, will the resulting compound be neutral? Explain.

    Yes ______ No ______

    Write the chemical formula for this compound and explain what it means.

78. Can K ions and Cl ions form bonds? Explain your answer.

    Yes ______ No ______

79. Predict the number of K and Cl ions that will combine to form a neutral compound. ______ Write a chemical formula for this compound. ______ What is the name of this compound? ______

80. The net charges on the magnesium and oxide ions are ______. Can these two ions form an ionic bond?

    Yes ______ No ______

81. The chemical formula for magnesium oxide is ____________. If this compound were decomposed into its elements, there would be ____________ atom(s) of oxygen produced for every 1 atom of magnesium.

82. In formulas for compounds containing ionic bonds, the ____________ ion is listed first, and the ____________ ion is listed second.

In the examples considered previously, the net charges on the positive and negative ions were the same. In such cases, neutral compounds are formed when the ions combine in a 1:1 ratio. When the positive and negative ions have different net charges, they will combine in a different ratio to form neutral compounds.

Questions

Refer to the table of ions in question 71 as you answer the following questions.

83. The net charge on the Ca ion is ______. The net charge on the Cl ion is ______. These two ions can form ionic bonds because __________________.

84. One Ca ion contributes ______ positive charges. One Cl ion contributes ______ negative charge. How many chloride ions are required to equal the charge on each calcium ion? ___________

85. Write the chemical formula for the compound that forms between Ca and Cl and explain what it means.

    The name of this compound is __________________.

86. If this compound were decomposed, it would release ______ atom(s) of Cl for every one atom of calcium.

87. The ions within calcium chloride are held together by ___________. Ionic bonds are the attractive force between ions of ___________ sign.

88. The charge on an Al ion is ______, and the charge on a Cl ion is ______. Predict the formula of a compound that can form using these ions. ___________

89. Look up the charges for the oxide ion and the aluminum ion. Would you predict that these ions could bond? Explain.

    Yes ______ No ______

90.*Predict the formula for the compound that contains aluminum and oxide ions. (Hint: more than one ion of each type will be needed.)

Ionic Bonding: An Overview

The first phase of our survey of the elements has now been completed. You have a guideline, the octet rule, that you can use to predict which elements will be stable and which will react. Elements that have an octet of electrons in the valence shell are stable. Those that do not have a valence octet will tend to create one during chemical reactions. The cases we have considered in this chapter include those elements that can form a valence octet by either gaining or losing one to three electrons. Elements will gain an electron or two if that will complete an octet. Other elements may empty a valence shell by losing one to three electrons and thus expose a full octet in the shell below.

Elements that have gained or lost electrons are no longer neutral. Opposites attract, at least in the world of electric charges. When the negatively charged anions are close enough to the positively charged cations, ionic bonds form. Many compounds that contain ionic bonds form crystals and are quite stable. Common table salt, NaCl, is an example.

What is the world really like in this compound? Let's take our imaginary spaceship into the ultramicroscopic world once again. As we approach the salt crystal, the landscape is stark and craggy. Steep cliffs, crevasses, and imposing pillars are all sights we encounter. As we move in closer and smaller, we begin to see that the solid cliff face appears to be constructed of tiny bubbles of two different sizes. The smaller bubbles are the electron clouds of the sodium ions, with an estimated diameter of 0.19 nanometers. The larger structures are the electron clouds of chloride ions, with an estimated diameter of 0.36 nanometers. The chloride is larger because it is using three electron shells, whereas the smaller sodium ion has electrons in only the first two shells. Each negative chloride ion is surrounded on all sides by small, but positive, sodium ions. Conversely, the positive charge on a sodium ion is attracted to the negative charges of chloride ions found top and bottom, front and back, right and left. Within the crystal, we find a lattice, represented in the diagram on the following page.

If we continued our journey into the electron clouds, smaller and smaller, we would again find tracts of empty space before finally discovering the tiny, extremely dense, and positively charged nuclei of the sodium and chloride ions. Think of this tiny world the next time you shake the white crystals on your food and appreciate its savor.

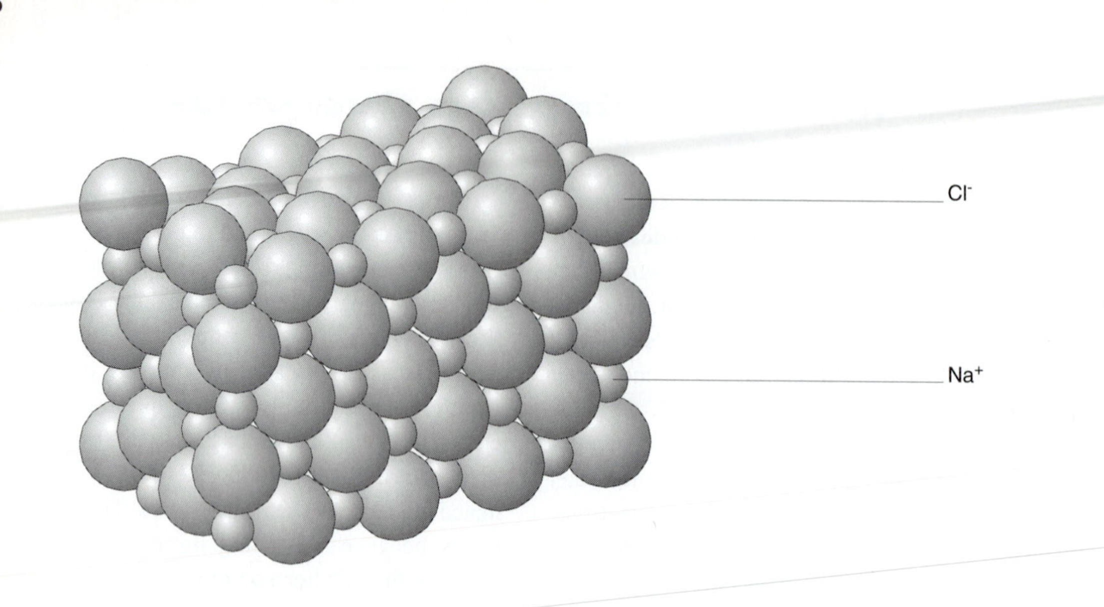

A Sodium Chloride Crystal

Answers

1. Yes.

   Water is a compound, for the following reasons. The question states that water can be decomposed by an electrical current. This suggests that water is a compound, because a compound can be decomposed into simpler substances during a chemical reaction. The problem also states that hydrogen and oxygen are produced in a 2:1 ratio. This verifies that water is a compound, because elements are present in definite proportions in compounds.

2. Chemical bonds; molecules

3. No.

   Water is a compound that contains hydrogen and oxygen in a 2:1 ratio. By definition, a compound contains elements in definite proportions. The substance in this question released hydrogen and oxygen in equal amounts. (This substance is a different compound, has different properties, and is called hydrogen peroxide.)

4. No.

   Saltwater is a mixture, not a compound. You can mix the salt and water in any proportion you like. The substances are not elements and are not present in definite proportions. Thus, saltwater is not a compound.

5. 4 elements; 1 sodium; 1 hydrogen; 1 carbon; 3 oxygens.

6. 6 atoms in each molecule. The atoms are linked by chemical bonds.

7.

Formula	Element or Compound?
H*	Element
CO	Compound
H_2O	Compound
O*	Element
$C_6H_{12}O_6$	Compound

Two formulas represent elements (H and O). Each lists only one atom. Since compounds must contain two or more elements in definite proportions, these cannot be compounds. All the structures that are compounds contain more than one element. The kinds of atoms are represented by chemical symbols. The number of each kind of atom is indicated by the subscript that follows it. When elements are listed together (no spaces), it indicates that atoms of these elements are components of a compound, associated by chemical bonds.

8. First (n = 1) electron shell; two electrons; two electrons; yes

9. Electron shells 1 and 2 (n = 1 and n = 2); eight electrons; eight electrons; no

10. Electron shells 1, 2, and 3 for argon; electron shells 1, 2, 3, and 4 for krypton; electron shells 1 through 5 for xenon.

11. The outermost shell occupied in argon is principal energy level 3. It contains eight electrons. The outermost shell occupied in krypton is 4. It contains eight electrons. The outermost shell occupied in xenon is 5. It contains eight electrons.

*Oxygen and hydrogen are not usually found in the atomic or elemental form in nature, because these forms are not stable. Both form diatomic molecules, O_2 and H_2. The formulas indicate that two atoms of oxygen or two atoms of hydrogen bond together to form each diatomic molecule. In these molecules, the atoms are linked by covalent bonds, a form of bonding that is covered in Chapter 3.

12.

Element	Symbol	Valence Shell	# Valence Electrons	Stable or Unstable
Helium	He	1	2	Stable
Neon	Ne	2	8	Stable
Argon	Ar	3	8	Stable
Krypton	Kr	4	8	Stable
Oxygen	O	2	6	Unstable
Hydrogen	H	1	1	Unstable
Magnesium	Mg	3	2	Unstable
Nitrogen	N	2	5	Unstable
Carbon	C	2	4	Unstable

Helium, neon, argon, and krypton are all noble gases, thus stable and unreactive. All except helium have eight electrons in the valence shell. Thus, stability is correlated with the presence of eight valence electrons, as stated in the octet rule. Small elements, such as helium, that use only the first electron shell are stable, with just two valence electrons. All the other elements have a different number of valence electrons, tending to be less stable and more reactive than the noble gases.

13. 8

14. Valence shell is 6; no.

    Radon would not be chemically reactive, because it contains an octet of electrons in its valence shell ($n = 6$, the outermost shell occupied by electrons). An octet of valence electrons is a very stable structure.

15. 8

16. One; no. Information provided about these elements states that they are unstable and undergo vigorous chemical reactions.

17. These elements are located at the far left in the periodic table. They are arranged in a column labeled "IA."

18. Cesium is just below rubidium (Rb) in the group IA column of elements.

19. One; no.

    All the elements in the group IA column have one valence electron. Cs would not be a stable element like the noble gases. It does not have an octet of electrons in its valence shell. It has only one valence electron, and this configuration is associated with reactivity.

20. Stable; unstable

21. Atomic number is 11; 11 protons; 11 electrons; 0 overall charge

22. Electron; ionization

23. 11 protons; 10 electrons; net charge is +1; ion

24.

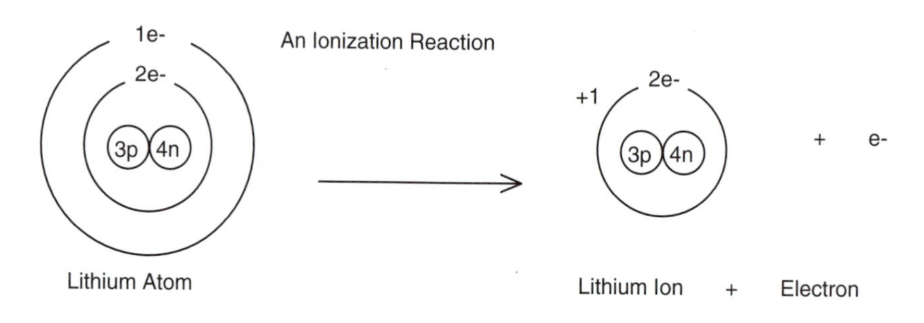

25. Lithium ion; +1

26. Potassium ion; +1

27. Two valence electrons; no.

    These elements do not have an octet of valence electrons. An octet of electrons in the valence shell is the configuration that is associated with stability. Also, information provided about these elements states that they are reactive.

28. The four elements listed above are arranged in a column that is located at the left in the periodic table. This group (group IIA) is just to the right of group IA and is the second column from the left.

29.

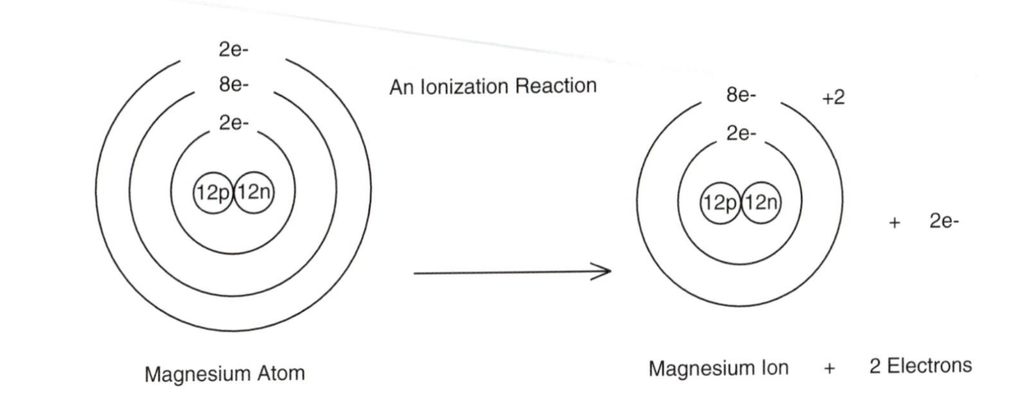

30. Atomic number is 12; 12 protons; 12 electrons; 0 overall charge

31. 12 protons; 10 electrons; net charge is +2

32. Atomic number is 13; symbol is Al; valence shell is 3

33.

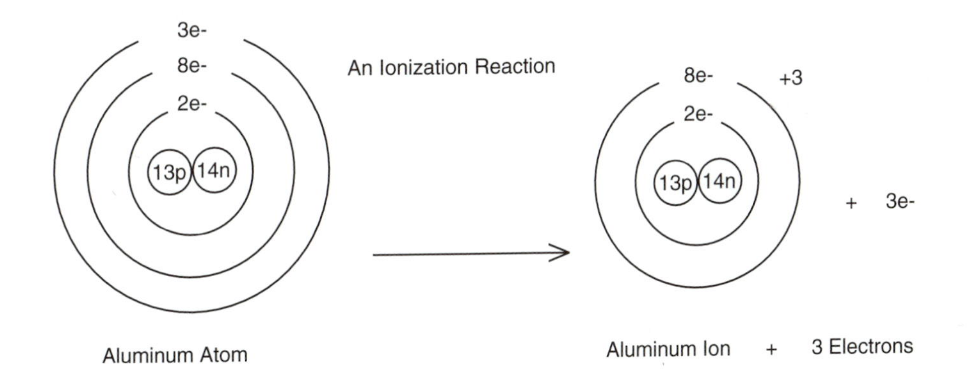

34. 13 electrons; 13 protons; 14 neutrons; electron charge is –1; proton charge is +1; neutron charge is 0.

35. After ionization: 10 electrons; 13 protons; 14 neutrons; net charge is +3

(The 10 electrons contribute ten negative charges. The 13 protons contribute 13 positive charges. The algebraic sum is +3.) Also called a cation; moves to negative terminal or cathode.

36.

Element	Atom		Ion	
	Valence Shell	# Valence Electrons	Valence Shell	# Valence Electrons
Li	2	1	1	2
Na	3	1	2	8
Mg	3	2	2	8
Al	3	3	2	8
K	4	1	3	8
Ca	4	2	3	8
Rb	5	1	4	8
Sr	5	2	4	8

37. One, two, or three electrons before reacting; eight after reacting,

    (or two in the case of Li, a small element that uses only the first electron shell in the ion).

38. The octet rule states that elements that have eight electrons in the valence shell (or two if only the first electronic shell is involved) are stable. Elements that have one, two, or three valence electrons are not stable. Such elements will lose their valence electrons and form cations (positive ions). This exposes an octet of electrons in the electron shell below, and it now becomes the valence shell. Because the valence shell in the ion has an octet of electrons, the ion is stable.

39. These elements are arranged in a column and are members of group VIIA. This column is at the right-hand part of the periodic table, just to the left of the noble gases (group VIIIA).

40. Seven valence electrons; gains (or accepts) one electron; stable.

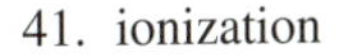
41. ionization

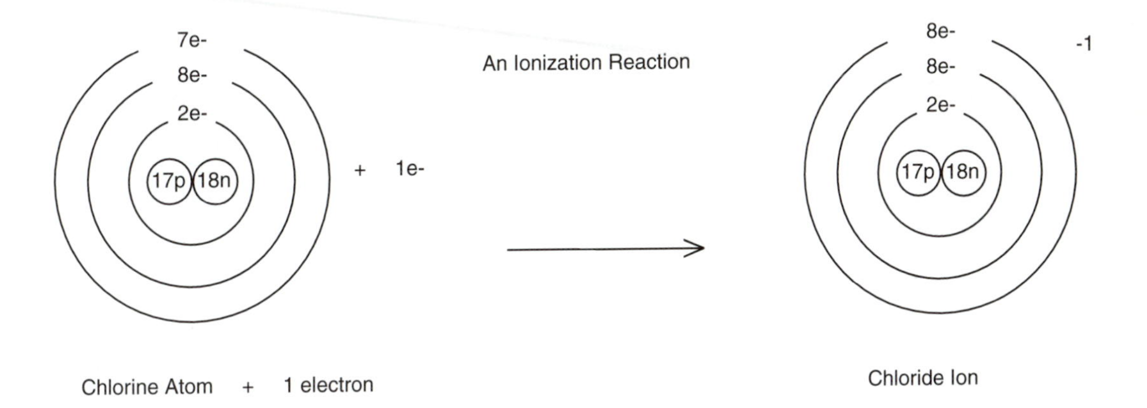

42. After ionization: 17 protons; 18 electrons; net charge is –1.

Each proton has a charge of +1. Since there are 17 protons, the total charge from protons is +17. There are 18 electrons, each with a charge of –1, for a total of –18. The sum of +17 and –18 is –1.

43. fluoride

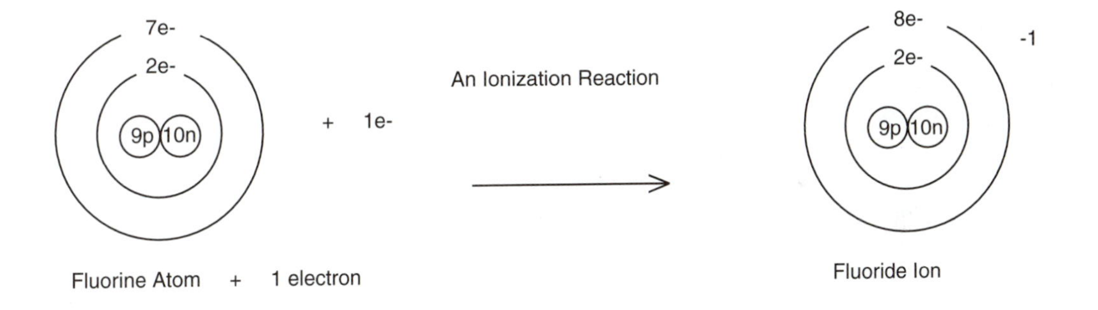

44. Net charge for fluoride is –1; net charge for chloride is –1; net charge for bromide would be –1.

45. One electron; produces anions; net charge of –1.

46. These elements are located in a column at the right-hand part of the periodic table and to the left of group VIIA. The group containing O, S, and Se is called group VIA.

47. Elements in group VIA can acquire an octet of valence electrons by accepting two electrons.

48.

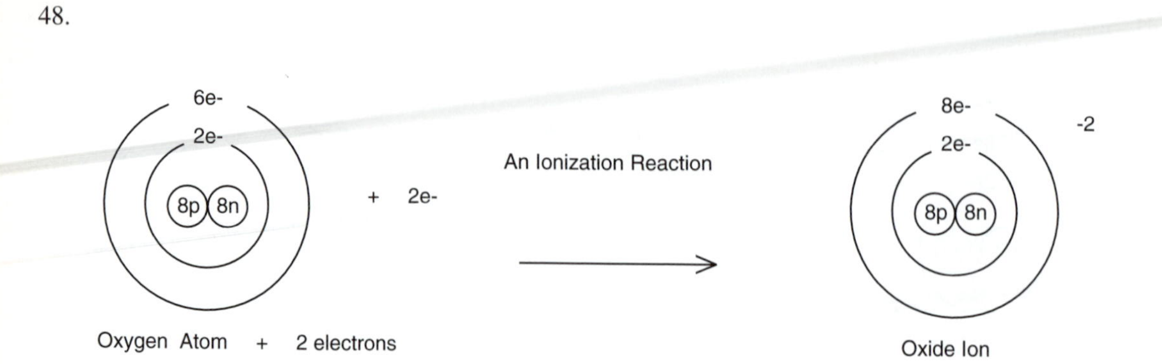

49. Oxygen atom: eight protons; eight electrons; net charge of 0. Oxide ion: eight protons; ten electrons; net charge of –2.

50. Gain; ionization; –2.

51. Lose; cations; oxidation.

52. Elements with one, two, or three valence electrons tend to be listed in the left-hand part of the periodic table.

53. Donors; oxidized.

54. Gain; octet; reduction.

55. Acceptors; reduced.

56. These elements are listed in the left-hand part of the periodic table.

57. These elements are listed in the right-hand part of the periodic table.

58. Oxidized; reduced.

59. Column; VIIA; right-hand part.

60. Column; IA; far left.

61. Elements that have the same number of valence electrons are grouped in a column. The name of the group includes a number (in Roman numerals) that gives the number of valence electrons. For example, elements with four valence electrons are arranged in a column labeled IVA.

62. Electrons participated in chemical reactions. Neither protons nor neutrons were involved.

63. Three occupied shells; valence shell electrons; oxidation.

64. Three occupied shells; valence electrons; reduction.

65. Valence electrons participate in chemical reactions. Neither protons, neutrons, nor inner-shell electrons were involved in the simple ionization reactions considered here.

66. Valence electrons; checking the number present in the name of a group in the periodic table.

67. The octet rule predicts that: an element containing eight electrons in the valence shell will be stable and unreactive, an element containing six or seven valence electrons will accept electrons to complete a valence octet, and an element containing one to three valence electrons will lose the valence electrons during chemical reactions, exposing an octet of electrons in the shell below.

68.

Element	# Valence Electrons	Type of Reaction
C	4	Shares electrons
Na	1	Loses an electron—oxidation
Ca	2	Loses two electrons—oxidation
O	6	Gains two electrons—reduction
N	5	Shares electrons
Mg	2	Loses two electrons—oxidation

Na, Ca, and Mg are all metals with one or two valence electrons. These elements lose the valence electrons and become oxidized. This exposes a full octet of electrons in the shell below, and the resulting ion is stable. Oxygen is a nonmetal with six valence electrons. It accepts two electrons to complete an octet in its valence shell and achieve stability. A gain of electrons is reduction. Carbon has four valence electrons, and nitrogen has five. These elements share electrons to achieve stability.

69. Lose electrons; cation; +2; oxidation; metal; toward the left.

70. Gain electrons; anion; –2; reduction; nonmetal; toward the right.

71.

Element	# Valence Electrons	Ionization	Stable Ionic Form
Na	1	Yes	Na^{+}
C	4	No	None
K	1	Yes	K^{+}
Cl	7	Yes	Cl^{-}
H	1	Yes	H^{+}
Mg	2	Yes	Mg^{2+}
O	6	Yes	O^{2-}
Al	3	Yes	Al^{3+}
N	5	No	None

72. No. Na and K cannot form ionic bonds, because they both form positive ions. Two positive ions will repel each other.

73. Yes. Na forms a positive ion, and Cl forms a negative ion. Ions with opposite charges are attracted to each other and form ionic bonds.

74. No. Na forms positive ions. Carbon has four valence electrons and does not form any type of ion. Because one of the elements does not ionize, no ionic bonding is possible.

75. Attracted; ionic; repel; cannot.

76. +1; –1. Yes. These ions have opposite charges. Ions of opposite charge are attracted to each other and form ionic bonds.

77. Yes; Na has one positive charge. Cl has one negative charge. The sum of these charges is zero. The compound is neutral because these charges are equal. NaCl is the formula for this compound. It shows that sodium ions and chloride ions are present in a 1:1 ratio in the sodium chloride compound. The ions are attracted to each other by ionic bonds.

    Recall that in ionically bonded compounds, individual molecules do not form. Rather, an ion is attracted to ions of the opposite sign on all sides (see the diagram on page 66, the chapter overview). A formula, such as NaCl, thus indicates that Na ions and Cl ions are present in equal numbers in the compound, but does not imply that one particular sodium ion is bonded to one particular chloride ion to form a molecule.

78. Yes. Potassium forms positive ions, and chloride forms negative ions. Ions with opposite charges are attracted to each other and can form ionic bonds.

79. One of each; KCl; potassium chloride.

80. +2 for magnesium and –2 for oxygen. Yes, they can bond.

81. MgO; 1 atom of oxygen produced for every 1 atom of magnesium.

82. Positive ion first and negative ion second.

83. +2; –1; they have opposite charges, and opposite charges attract each other.

84. 2 positive charges; 1 negative charge. Two chloride ions are required for each calcium ion.

   ($2 \times -1 = -2$). Because –2 and +2 are equal in magnitude and opposite in sign, this combination will be neutral.

85. $CaCl_2$ is the formula for this compound, which indicates that there are two chloride ions for every one calcium ion. The name of this compound is calcium chloride.

86. 2

87. Ionic bonds; opposite.

88. +3; –1; $AlCl_3$.

89. Charge on the oxide ion would be –2; the charge for the aluminum ion would be +3; these ions could bond because they have opposite charges.

90.*The formula for the compound that contains aluminum and oxide ions is Al_2O_3.

   The subscript 2 means there are two aluminum ions. Each has a charge of +3. Both aluminum ions together have a total of six positive charges. The subscript 3 means there are three oxide ions. Each oxygen ion has a charge of –2. Since $-2 \times 3 = -6$, the oxygen ions contribute exactly the amount of charge necessary to equal the aluminum ions.

Chapter Test: Ionic Bonding

The questions in this chapter test evaluate your mastery of all the objectives for this unit. While material from Chapter 1 is not tested directly, recall of the basic material from that chapter is necessary background in some cases. Take this test without looking up material in the chapter. You may, however, use the periodic table. Questions that test intermediate objectives are indicated by an asterisk, and you may omit them if you did not cover that material in the chapter. After you have completed the test, you can check your work by using the answer key located at the end of the test.

1. A molecule represented by the formula C_3H_8O contains ______ atom(s) of carbon, ______ atom(s) of hydrogen, ______ atom(s) of oxygen, and ____________ atom(s) of nitrogen.

2. Define the term "compound." ____

   ____

3. The formula NaBr represents a(n)

   compound ______ element ______

   Explain:

   ____

4. The element Cl has an atomic number of ____________. The valence shell in chlorine is ____________, and it contains ______ electrons. Chlorine is in group ______. Elements in this group are usually

   stable and unreactive ______ unstable and reactive _______

   Explain your answer.

   ____

   ____

5. State the octet rule in your own words. Explain how the octet rule can be used to predict which elements are stable and which are not.

   ____

   ____

   ____

6. Complete the table below by filling in the missing information. Explain your answers.

Element	Symbol	Valence Shell	# Valence Electrons	Stable or Unstable
Neon	Ne	2	8	Stable
Carbon				
Oxygen				
Calcium				
Hydrogen				

____

____

____

7. Potassium is a member of group ______. It has ______ valence electron. During chemical reactions, this element tends to ____________.

8. In the space below, diagram the chemical reaction that would be predicted for potassium atoms. After a reaction, the structure that has formed is a ____________ with a net charge of ____________.

9. The element potassium is a

   metal ______ nonmetal ______

10. Potassium normally undergoes

    oxidation ______ reduction ______

11. Elements that have ____________ valence electrons tend to form cations. Explain why the cations that form are stable.

____

____

12. Oxygen is an element that is

    stable ______ unstable ______

    It has ______ valence electrons. Elements with this number of valence electrons tend to be

    metals ______ nonmetals ______

13. During a chemical reaction, oxygen tends to ____________ electrons. After such a reaction, the structure that has formed is called an ____________, and it has a net charge of ____________. A reaction of this type is an example of

    oxidation ______ reduction ______

14. State a general rule that can be used to predict which elements will form negative ions.

    ____

15. Explain why the anions that form from the elements described in question 14 are stable.

    ____

16. Look up the element chlorine in the periodic table. It is a member of what group? ________________

    How are the members of this group arranged in the periodic table?

    ____

17. In the space below, sketch the ionization reaction that would be predicted for a chlorine atom. Label the diagram.

18. Chlorine ionization is an example of

oxidation ______ reduction ______

19. Would you predict that chlorine and potassium could form ionic bonds? Explain.

Yes ______ No ______

____

20. If you predicted that chlorine and potassium could form ionic bonds (see question 19), write the formula for the product that would be formed. What is the name of this substance?

____

21. What ions form from the elements calcium and bromine? ____________ Could these elements form ionic bonds with each other?

Yes ______ No ______

22. If Ca and Br do react (see question 21), state the formula for the product that is formed. Explain what the formula means.

____

23. Pairs of elements are listed below. Work out the charge for the ion that would form from each element. Combine the ions in the correct proportions to form a neutral compound. Give the formula for that compound.

Elements	Positive Ion	Negative Ion	# Ions in Compound	Formula of Compound
Mg and F	Mg^{2+}	F^{-}	1 Mg (+2) and 2 F (–2)	MgF_2
Al and Br				
Mg and Cl				
Na and O				
Al and O*				

24.*In a sample of solid NaCl, does one Na^{+} ion bond with one or many Cl^{-} ions? Explain your answer.

____

____

Answers

1. 3 atoms of carbon; 8 atoms of hydrogen; 1 atom of oxygen; no atoms of nitrogen.

2. A compound is a substance that contains two or more elements combined in a definite proportion by mass. A compound can be decomposed into simpler substances (the elements in it) by chemical reactions.

3. Compound.

   The formula for a compound lists all the elements that are present in the compound and uses subscripts to show the proportions of each element. NaBr is such a formula. It means that sodium and bromine are present in a 1:1 ratio, producing a compound called sodium bromide. Since NaBr contains more than one element in a definite ratio, NaBr fits the definition of a compound. If the Na and Br did not bond and form a compound, each would be listed separately.

4. Chlorine's atomic number is 17; valence shell is 3 ($n = 3$); seven electrons; group VIIA; unstable and reactive.

   Chlorine is unstable because it has seven valence electrons. It gains one electron during chemical reactions to achieve stability by completing an octet in the valence shell. An octet of valence electrons is stable.

5. The octet rule states that elements containing eight electrons in the valence shell are stable and unreactive. Elements that do not have an octet of electrons are generally less stable and more reactive. If an element does not have an octet of electrons initially, it may go through a chemical reaction, creating a valence octet.

   To predict whether an element is stable or not, look up the element in the periodic table. The name of the group to which the element belongs (at the top of the column) contains a Roman numeral. This Roman numeral is the number of valence electrons for elements in that column. If there are eight valence electrons, predict that the element is stable. If there are not eight electrons in the valence shell, predict that the element is unstable. Small elements such as helium are stable with two valence electrons, because only the first electronic shell is used.

6. Complete the table below by filling in the missing information.

Element	Symbol	Valence Shell	# Valence Electrons	Stable or Unstable
Neon	Ne	2	8	Stable
Carbon	C	2	4	Unstable
Oxygen	O	2	6	Unstable
Calcium	Ca	4	2	Unstable
Hydrogen	H	1	1	Unstable

Neon is stable and unreactive because it has an octet of electrons in the valence shell. All the other elements are unstable and reactive, because they do not have an octet of valence electrons.

7. IA; one valence electron; lose the one electron in its valence shell.

8. Potassium ion; +1 charge.

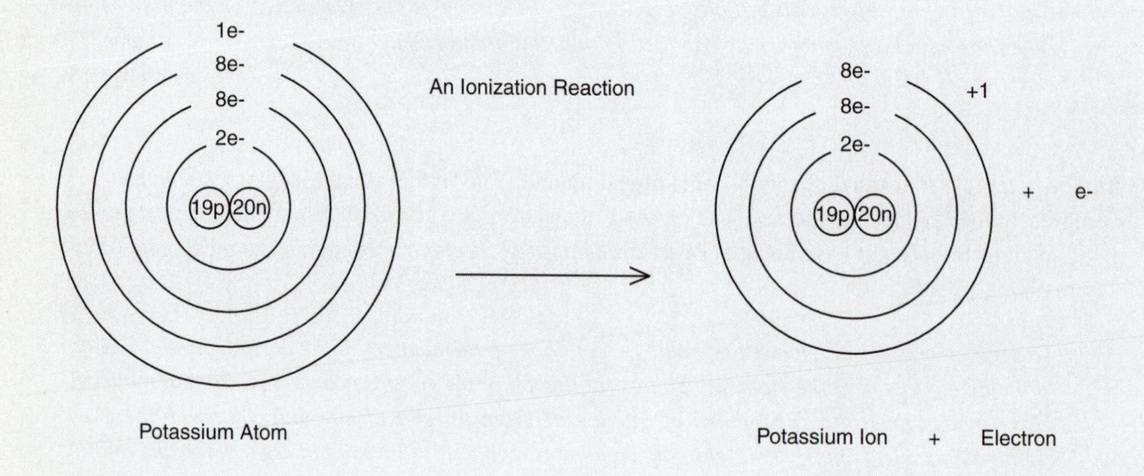

9. Metal.

10. Oxidation.

11. 1 to 3.

    Elements that have one to three valence electrons tend to lose their valence electrons during chemical reactions. This empties the original valence shell and exposes the electron shell beneath it. The newly exposed shell becomes the valence shell. This shell contains an octet of electrons. Because an octet of electrons in the valence shell is stable, the ions formed in this way are stable.

12. Unstable; six valence electrons; nonmetals.

13. Gain two; anion; –2; reduction.

14. Elements that have six or seven valence electrons will gain electrons and form negative ions.

15. Elements with six valence electrons will accept two more electrons. Elements with seven valence electrons will accept one more electron. In both cases, this completes an octet of electrons in the valence shell. This is a stable configuration. The ions are negatively charged, because there are now more electrons than protons.

16. VIIA; They are arranged in a column in the right-hand part of the periodic table and to the left of the noble gases.

17.

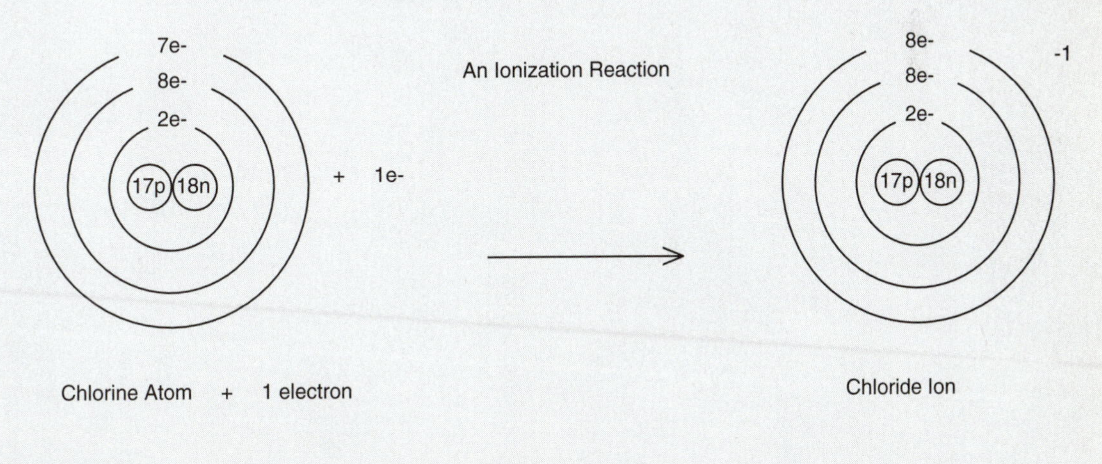

18. Reduction.

19. Yes; Chlorine forms a negative ion, and potassium forms a positive ion. Ions of opposite sign are attracted to each other, thus forming ionic bonds.

20. The formula is KCl. The compound is potassium chloride.

21. Ca^{2+} and Br^-, Yes.

22. The product would have the formula $CaBr_2$. This formula means that there is one atom of calcium bonded to two atoms of bromine in the compound.

23.

Elements	Positive Ion	Negative Ion	# Ions in Compound	Formula of Compound
Mg and F	Mg^{2+}	F^-	1 Mg (+2) and 2 F (–2)	MgF_2
Al and Br	Al^{3+}	Br^-	1 Al (+3) and 3 Br (–3)	$AlBr_3$
Mg and Cl	Mg^{2+}	Cl^-	1 Mg (+2) and 2 Cl (–2)	$MgCl_2$
Na and O	Na^+	O^{2-}	2 Na (+2) and 1 O (–2)	Na_2O
Al and O*	Al^{3+}	O^{2-}	2 Al (+6) and 3 O (–6)	Al_2O_3

24.*In solid NaCl, one sodium ion bonds to many chloride ions. In a crystal, every ion is surrounded by many ions with an opposite charge. Every ion is attracted to ions of opposite charge on all sides. All these attractions are ionic bonds. The figure on page 66 shows this arrangement.

Self-Evaluation

Compare your answers on the chapter test to the correct answers given in the answer key. If you answered most questions correctly, you may be ready to proceed to the next chapter. Your own goals should be used to determine this. If you missed questions on a topic that is important to your academic program, then you should review that topic before continuing. Repeat the sections of the chapter that explain and test the topic you had difficulty mastering; then try the chapter test again. Repetition of this chapter and its questions is a good way to reinforce your learning and will help you remember the new material.

Chapter 3

Covalent Bonding

Objectives

1. Define the following terms and apply the definitions correctly: covalent bond, double bond, triple bond, electron dot structure, molecule, electronegativity, polar covalent bond, reactants, products, and balanced equations.
2. Use the octet rule to predict which elements always form covalent bonds.
3. Predict which elements form covalent bonds in some situations and ionic bonds in others. Use the octet rule to predict what type of bond will form in a given situation.
4. Use electron dot symbols to represent atoms.
5. Use electron dot symbols to represent the structure of simple molecules that contain atoms joined by covalent bonds.
6. Use chemical formulas and structural formulas to represent molecules.
7. *Use the electronegativity scale to predict which combinations of elements will form polar covalent bonds.
8. Represent chemical reactions using equations.
9. * or ** Balance equations that represent chemical reactions.

   Some of the questions on this topic are intermediate or advanced in difficulty.

The Octet Rule Revisited

Elements that have eight valence electrons are stable. Elements with one, two, or three valence electrons are unstable and achieve stability by losing their valence electrons, thus exposing an octet in the shell below. Elements with six or seven valence electrons are unstable. They may achieve stability by gaining enough electrons to complete an octet in the valence shell or they may achieve stability by sharing electrons. Elements with four or five valence electrons always share electrons to create an octet in the valence shell.

Questions

Use the octet rule and the periodic table of the elements to help you answer these questions.

1. Fill in the missing information in the table below. The first example is worked and should be used as a model for the remaining questions.

Element	# Valence Electrons	Stable or Unstable	Type of Chemical Reaction Predicted by the Octet Rule
Oxygen	6	Unstable	May accept two electrons to complete valence octet or shares two electrons
Chlorine			
Carbon			
Neon			
Nitrogen			
Hydrogen			
Calcium			

2. Elements with ____________ valence electrons are unstable and always achieve stability by sharing electrons. Elements with ____________ valence electrons are unstable and may accept electrons to complete an octet, or may share electrons to achieve stability.

3. The element chlorine can ionize by ____________ one electron. Cl can function as an electron acceptor only if another element is available to ____________ electrons. Elements that can function as electron donors are those with ______ valence electrons.

4. Chlorine cannot accept electrons if no element is available to ____________ them. If Cl cannot acquire electrons by ionization, then it may ____________ electrons to complete its octet. Elements with ____________ valence electrons always achieve stability by sharing, and those with ______ valence electrons may achieve stability by sharing. Therefore, Cl is likely to share electrons when it reacts with an element that has ____________ valence electrons.

The Covalent Bond

Atoms that achieve stability by sharing electrons may share two, four, or six electrons. A single covalent bond is one pair of electrons shared between the two participating atoms. Two shared pairs (four electrons) is a double bond, and three shared pairs (six electrons) is a triple bond. The shared electrons belong to both the sharing atoms simultaneously. The shared electrons help complete a valence octet for each of the participating atoms.

Questions

5. Carbon atoms can bond to many other elements by ____________ electrons. Sometimes carbon atoms bond to each other. If two carbon atoms share two electrons between them, this is ____________ pair of electrons and is an example of a ____________ covalent bond.

6. Oxygen is an element that has ____________ valence electrons. It can ionize by ____________ two electrons when an electron ____________ is available. Give two examples of elements that serve as electron donors.

7. If two oxygen atoms react with each other, would you predict that they will ionize? If not, what will they do? Explain your answer.

   Yes ______ No ______

8. Write a rule that can be used to predict when an element with six or seven valence electrons will form ionic bonds and when it will share electrons to achieve stability. (Consider your answers to questions 4, 6, and 7 as you work on this.)

9. Nitrogen is an element that has ______ valence electrons. It tends to ______ electrons to complete its octet. When two nitrogen atoms bond with each other, six electrons are shared between them. There are ______ shared pairs of electrons or a __________________ bond.

10. Two shared pairs of electrons form a ____________ bond. The attraction between ions of opposite sign is an ____________ bond. One shared pair of electrons is a ____________ bond. Three shared pairs of electrons is a ____________ bond.

11. Atoms that do not have an octet of valence electrons are ____________. Such atoms can achieve stability by acquiring an octet in the valence shell. This can be done by __________________, __________________, or __________________ electrons.

Representing atoms with a modified type of Lewis dot structure is often a useful exercise. The symbols can be used to work out the structures that form in compounds containing covalent bonds. Examples of modified Lewis electron dot structures are shown below. The examples include hydrogen, carbon, nitrogen, and chlorine.

·C· H· :N· :Cl·

Questions

12. What is the atomic number of hydrogen? ______ Hydrogen contains ______ valence electron(s). There is ______ dot in the electron dot representation of hydrogen.

13. What is the atomic number of carbon? ____________ How many valence electrons does carbon contain? ____________ What is the chemical symbol for carbon? ____________ In the electron dot representation of carbon (see above), how many dots are arrayed around its symbol? ____________

14. Nitrogen has ______ valence electrons, and its electron dot symbol has ______ dots.

15. Based on the pattern you have observed, state a general rule for writing the electron dot symbol for an element.

16. Follow the guidelines you have developed, and write electron dot structures for the elements listed in the table below.

Element	# Valence Electrons	Electron Dot Structure
Oxygen		
Neon		
Fluorine		

Methane (the major component of natural gas) is an example of a compound that contains atoms linked together by covalent bonds. Atoms that are joined together by covalent bonds form a molecule. A molecule is the smallest particle of a compound that still retains the properties of the compound. The formula for methane is CH_4. The electron dot structures for the five participating atoms are written below, on the left. The arrangement of atoms in the methane molecule is shown on the right.

$$\cdot\dot{\underset{\cdot}{C}}\cdot \qquad H\cdot \qquad H\cdot \qquad H\cdot \qquad H\cdot \qquad\qquad \begin{array}{c} H \\ H:\overset{..}{\underset{..}{C}}:H \\ H \end{array}$$

Questions

17. Before the reaction that forms methane has occurred, there are ______ separate atoms. After methane has formed, there is one ______ that contains the ______ atoms now linked into one structure by ____________.

18. The smallest particle of methane that still has the properties of methane is a ___________ of methane.

19. The answers to this question should consider only valence electrons shown in the electron dot structures. The carbon atom contributes ____________ electrons, and each hydrogen contributes ____________ electron to the methane structure. The total number of valence electrons in the five contributing atoms is ____________. There are ____________ electrons shown in the methane structure.

20. How many electrons are arranged around the carbon atom in the methane molecule? ____________ The carbon atom shares all these electrons. Based on this information, would you predict that this structure is stable? Explain your answer.

21. How many electrons are next to each hydrogen in the methane molecule? ____________. The electrons that are next to a hydrogen are shared by it. Based on this information, would you predict that this structure is stable? Explain your answer.

22. A pair of electrons that is between two atoms belongs to both atoms; the two atoms share the electrons. A shared pair of electrons is a ____________. Do shared electrons satisfy the octet rule requirements of both the sharing atoms at the same time? Explain your answer.

23. Hydrogen has only ____________ electron and therefore has no complete inner shells. Carbon has ____________ complete inner shell(s) that contain(s) ____________ electrons. Do these inner-shell electrons participate in the chemical bonding that forms methane?

Another way of representing the methane molecule is shown below. This is a structural formula. In structural formulas, each atom is represented by its chemical symbol, and each covalent bond is represented with a dash.

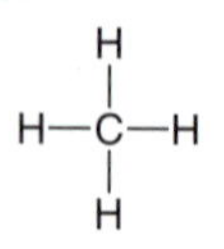

Questions

24. There are ____________ single covalent bonds in the methane molecule. Each bond consists of ____________ shared pair of electrons and is represented by ____________ dash.

25. If a pair of atoms shares four electrons, they would be linked by a ____________ covalent bond. This would be represented in a structural formula by ____________ dashes.

26. There is a compound called ethane with the chemical formula C_2H_6. How many atoms are present in the ethane molecule? ____________ Each carbon contributes ____________ valence electrons to the ethane molecule. Each hydrogen contributes ____________ valence electron(s) to the ethane structure. In all, there are ____________ electrons involved in forming the ethane structure.

27. In the space below, write electron dot structures for all the carbons and hydrogens that are components of the ethane structure. Write them as separated atoms.

28. For stability, carbon must share in an ____________ of electrons. Each hydrogen must share ____________ electrons to become stable.

29. In the space below, write an electron dot structure for the ethane molecule. Use the component atoms from question 27. Remember the octet rule requirements you stated in question 28. Show how the ethane molecule would be represented as a structural formula.

30. Electron dot structures show only valence electrons. Explain why these types of symbols are useful in working out the structures of molecules that contain covalent bonds.

A number of familiar and important compounds are composed of small molecules that contain atoms linked together by covalent bonds. Water, ammonia, oxygen gas (the form found in the air), and carbon dioxide are some examples. Several of these are important components of the earth's atmosphere, and all are important to living organisms.

Questions

31. Fill in the table below. First, write the electron dot structures of all the atoms contained in each molecule. Then work out the electron dot structure of the molecule. Remember to satisfy the octet rule requirements. Finally, write the structural formula for the compound.

Compound	Chemical Formula	Electron Dot Structure of Atoms	Electron Dot Structure of Molecule	Structural Formula
Water	H_2O			
Ammonia	NH_3			
Oxygen	O_2			
Carbon Dioxide	CO_2			

Note: Structural formulas show each covalent bond with a dash. One dash represents a single covalent bond (one shared pair of electrons). Valence electrons that are nonbonding are not shown.

32. How many electrons are around each oxygen atom in the molecules above? ____________ There are ____________ electrons around carbon and ____________ electrons around nitrogen. There are ____________ electrons shared by each hydrogen atom.

33. In light of your answers to question 32, write a general statement of the octet rule predictions for molecules containing covalent bonds.

Organic Chemistry—A Preview

Carbon is a very important element, because it is the basis of organic compounds. Organic molecules make up much of the structure of living organisms and are the major components in fossil fuels, which are derived from organisms that lived long ago. Organic molecules usually contain a substantial amount of hydrogen bonded to the carbon. Other elements may be present as well. The structures of a few organic compounds are shown below.

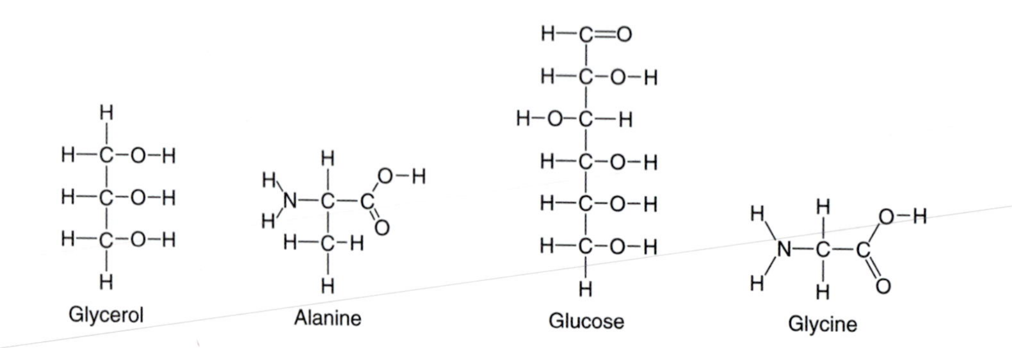

Questions

34. Hydrogen has ______ valence electron(s) and needs to share ____________ more electron(s) to create a stable pattern. Examine hydrogen in the molecules above. It always forms ____________ covalent bond(s).

35. Oxygen has ____________ valence electron(s) and needs to share ____________ more electron(s) to create a stable pattern. Examine oxygen in the molecules above. Oxygen always forms ____________ covalent bond(s).

36. Nitrogen has ____________ valence electron(s) and needs to share ____________ more electron(s) to form an octet in the valence shell. How many bonds does nitrogen form? (Refer to the molecules above to answer this question.) ____________

37. Carbon has ____________ valence electron(s) and needs to share ____________ more electron(s) to form a valence octet. Carbon forms ____________ covalent bond(s).

38. Fill in the table below.

	H	O	N	C
# Valence Electrons				
# Electrons Needed for Stability				
# Covalent Bonds Formed				

Polar and Nonpolar Covalent Bonds

Nonpolar covalent bonds are equal covalent bonds. They form when atoms of the same type share electrons. Because the two atoms are identical, they share a pair of electrons equally. In other words, the average location of the electron cloud is midway between the two atoms. Polar covalent bonds are unequal covalent bonds. Polar covalent bonds often form when two atoms of different types share electrons. The electron cloud is shifted toward the atom that attracts electrons more strongly. A polar covalent bond is the result of unequal sharing of electrons. An example of each type of bond is given below.

– C – C –	– C – O –
Nonpolar	Polar

Questions

39. A bond between two carbon atoms will be ____________, because ____________________________.

40. A bond between carbon and oxygen is polar, because the two atoms are ____________, and the electrons are shared ____________.

41. Oxygen is an atom that pulls electrons more strongly than carbon. The electron cloud is shifted a bit toward ____________ and away from ____________.

42. Electrons have a ____________ charge, so the electron cloud as a whole is also ____________ charged.

43. If the ____________ charged electron cloud is shifted slightly toward an atom, that atom will develop a slightly ____________ charge. In the bond between oxygen and carbon, the ____________ would develop a slightly negative charge.

44. In the bond between carbon and oxygen, the electrons are displaced ____________ from the carbon. Carbon thus develops a slightly ____________ charge. Explain why this is so.

45. Overall, the structure formed when carbon and oxygen bond is electrically neutral. However, the electron cloud is shifted ____________ oxygen and ____________ from carbon. A partial separation of charge is created. This creates a slightly ____________ pole near carbon and a slightly ____________ pole near oxygen. This is why the bond is described as ____________.

*Intermediate Difficulty

Electronegativity is a measure of the tendency of an atom to attract electrons. Electronegativity values for some of the atoms that are important in many organic structures are provided in the table below. The larger the number, the stronger the pull on electrons. When two atoms that have different electronegativities form covalent bonds, the bonds will be polar. The electron cloud shifts toward the atom with the greater electronegativity.

Atom	O	N	S	C	H	P
Electronegativity	3.5	3.0	2.6	2.5	2.1	2.1

Questions

46. The electronegativity of nitrogen is ____________, and the electronegativity of hydrogen is ____________. Since these values are different, a bond between these two atoms will be ____________. The electron cloud will shift toward ____________, because it is the more electronegative element.

47. Fill in the table below. The first example is worked for you.

A Bond Between Atom 1 and Atom 1	Type	Location of Electron Cloud
(1) Carbon Bonded to (2) Carbon	Nonpolar	Shared equally
(1) Carbon Bonded to (2) Hydrogen		
(1) Carbon Bonded to (2) Oxygen		
(1) Oxygen Bonded to (2) Oxygen		
(1) Oxygen Bonded to (2) Nitrogen		
(1) Sulfur Bonded to (2) Carbon		
(1) Nitrogen Bonded to (2) Nitrogen		

48. Use the periodic table of the elements to look up the elements listed above that have the highest electronegativities. Where are they located?

49. Fluorine (not listed in the chart above) has the highest electronegativity of any element. The electronegativity of chlorine is also quite high. Based on this information, what factors tend to cause an element to have a high electronegativity?

*Intermediate Difficulty

Ionic bonding and covalent bonding (nonpolar variety) represent two different ways of achieving stability. Ionic bonding is the result of electron transfer, and nonpolar covalent bonds are the result of equal sharing. Polar covalent bonds are an intermediate case. Polar covalent bonds can be described as the result of unequal sharing (a modified covalent bond) or of partial transfer (modified ionic bond). If a bond is only slightly polar, it is more like a covalent bond. Bonds that are more and more polar have more ionic character. Some examples are presented below.

H – H	δ^- – C – H δ^+	δ^- – O – H δ^+
Nonpolar	Slightly Polar	Strongly Polar

Questions

50. The electronegativity of H is ______. A bond between two hydrogens will be ____________. The average location of the electrons shared by the two hydrogens is ________________.

51. The electronegativity of H is ______, and that of C is ____________. The difference between the two values is ____________, which means that the bond linking them is ________________.

52. The Greek letter delta (δ) is a symbol for "partial." Look at the diagram of the bond between carbon and hydrogen. Which end of the structure is the slightly negative pole? ____________ Which end of the structure is the slightly positive pole? ____________ Explain your answers.

53. The electronegativity of O is ____________, and that of H is ____________. The difference between these two values is ____________, which means that the bond linking them is ________________.

54. Examine the diagram of the bond between oxygen and hydrogen. The negative pole is the ____________ end, and the positive pole is the ____________ end. Overall, this structure is electrically ____________. Although the overall structure is neutral, there is a partial internal separation of charge, because the electron cloud has shifted toward ____________. When this occurs, we say the bond is ____________.

55. The symbol ____________ is used to represent the negative pole in a polar covalent bond, and the symbol ____________ is used to represent the positive pole. The Greek letter δ is a symbol for ____________.

Equations for Chemical Reactions

Every chemical reaction can be represented by an equation. An example is provided below. The substances present before the reaction starts are called reactants. Their formulas are on the left. The substances produced during the reaction are called products. Their formulas are on the right. The chemical reaction is the process that converts reactants to products. It is represented by an arrow.

$$C + 2H_2 \rightarrow CH_4$$

Reactants → Products

Questions

56. These questions pertain to the reaction shown above. The reactants for this reaction are ____________ and ____________. The product of the reaction is ____________. The chemical reaction itself is represented by the ____________. Is it possible to tell whether the reaction will proceed rapidly or slowly?

    Yes ______ No ______

57. Write an equation to represent the reaction of the elements sodium and chlorine to form sodium chloride.

    The reactants for this reaction are ____________, and the product is ____________.

58. Consider the following equation: $Ca + Cl_2 \rightarrow CaCl_2$
    What are the reactants for this reaction? ____________ What is the product? ____________ Can you learn anything about the speed of this reaction from the equation?

    Yes ______ No ______

    What is Cl_2?

59. The ____________ of a chemical reaction are found on the right of the chemical equation and the ____________ are found on the left.

The equation below represents a more complex reaction than the one we considered previously. The number in front of an atom or molecule is called a coefficient. Coefficients tell us how many copies of the molecule are present. One is understood if no coefficient is given.

$$CH_4 + 2O_2 \rightarrow CO_2 + 2H_2O$$

↑

Coefficient

Questions

60. The reactants for the reaction above are ____________ and ____________. The formula for methane is ____________. The 4 in the formula for methane is called a ______. What does this number tell us? ________________________ How many molecules of methane are used in this reaction? ______

61. The formula $2O_2$ is used to indicate the second reactant. The smaller 2 is called a ____________, and it tells how many ____________ are present in one oxygen molecule. This type of molecule is called a ____________. This is the form of oxygen that is present in the air. It is quite common and fairly stable. The larger 2 is called a ____________. It indicates that there are ____________ oxygen molecules used in this reaction. The two oxygen molecules contain a total of ____________ oxygen atoms.

62. The products for the reaction above are __________________. The reaction produces ____________ molecule(s) of CO_2 and ____________ molecule(s) of H_2O.

63. Count the carbon atoms in the reactants for the above reaction. There is/are ______ carbon(s) in the reactants. There is/are ______ carbon(s) in the products. There is/are ______ hydrogen(s) in the reactants and ______ hydrogen(s) in the products. There is/are ______ oxygen(s) in the reactants and ______ oxygen(s) in the products.

64. Write a general statement summarizing your observations from question 63.

Balancing Equations

During chemical reactions, matter is conserved. In other words, the atoms that are present in the reactants are present in the products. Neither the number nor the kinds of atoms change. The atoms usually have changed partners. As the reaction occurs, bonds are broken in the reactants, and new bonds form as products are produced. The bonding pattern changes as the atoms change partners during the chemical reaction.

Questions

65. Water can be formed from hydrogen and oxygen. The formula for hydrogen is ______ (the diatomic molecule is the stable form that is generally available). The formula for oxygen is ______ (again, the diatomic molecule is the form that is generally available). The formula for water is ______. Write an equation for this reaction. Do not worry about the number of reactants and products yet.

66. An equation that lists the correct reactants and products but does not equalize the numbers of all kinds of atoms is not balanced. How many oxygen atoms are present in the reactants? ______ How many are present in the product? ______ How many molecules of water would be needed to make the oxygens on the product side equal to those on the reactant side? ______ Correct your equation to reflect this change.

67. Refer to the revised equation in question 66 to answer the following questions. There are now ______ hydrogens on the reactant side of the equation and ______ hydrogens on the product side. Since these numbers are not equal, the equation is not ________________. How many hydrogen molecules are required to supply the necessary number of hydrogen atoms? ______ This can be indicated by changing the number called the ____________.

68. Write the newly modified equation for the formation of water from oxygen and hydrogen. Describe what this equation means.

69. Is the equation in question 68 balanced now?

    Yes ______ No ______

    Explain what a balanced equation is.

70. In the reactants, oxygen is bonded to ____________. In the products, oxygen is bonded to ____________. In the reactants, hydrogen is bonded to ____________. In the products, hydrogen is bonded to ____________. Rewrite the equation from question 68. Replace all the chemical formulas with structural formulas. Put arrows on all bonds that break during the chemical reaction and a star on those that form.

*Intermediate Difficulty

Equations for chemical reactions should be balanced. When the equation is balanced, the number and kinds of atoms are the same on both sides of the equation. In other words, conservation of matter is reflected correctly in the balanced equation. The coefficients show what proportions of each reactant and product are used and produced during the chemical reaction.

Questions

71. Photosynthesis is a very important process that occurs in many organisms, including plants and algae. Carbon dioxide and water are the reactants for this process. Glucose and oxygen are the products. Write an equation to represent this metabolic pathway. Do not worry about balancing yet. Check page 96 if you do not remember the formula for glucose. The chemical formula can be obtained from the structural formula by counting the different kinds of atoms.

72. Refer to the equation in question 71 to help you answer these questions. How many carbons are present on the reactant side? ____________ There are ____________ carbons on the product side of the equation. Is this equation balanced as far as carbon is concerned?

    Yes ______ No ______

    A coefficient of ____________ is required in front of ____________ to achieve balance for carbon.

73. Rewrite the equation for photosynthesis including the correct coefficient for carbon.

    There are now ______ hydrogens in the reactants and ______ hydrogens in the products. Is the current form of the equation balanced for hydrogen?

    Yes ______ No ______

74. Rewrite the photosynthesis equation so that it is now balanced for hydrogen. Explain how you did this and why you did it this way.

75. Refer to the equation in question 74 to help you answer these questions. There are ______ oxygens in the reactant molecules and ______ in the product molecules. Is the equation now balanced for oxygen?

    Yes ______ No ______

76. Rewrite the photosynthesis equation so that it is balanced for all elements. Explain what the equation for photosynthesis means.

**Advanced Difficulty

A variety of methods can be used to balance equations. When the equation is simple, balancing can often be done by inspection. When the reaction is complex, a more systematic approach is often helpful. A method that tabulates the number and kinds of atoms before and after a chemical reaction is presented in the questions below.

Questions

77. The next set of questions will be based on this reaction: $KClO_3 \rightarrow KCl + O_2$. Complete the table below by filling in the missing information.

# of Atoms in Reactants	Element	# of Atoms in Products
1	K	1
	Cl	
	O	

78. The element ___________ is not balanced in the above equation. There are ___________ atoms of this element in the reactants (on the left) and ___________ atoms of this element in the products (on the right). Explain how you can equalize the number of atoms of this element on the two sides of the equation.

79. Rewrite the equation for this reaction with your new coefficients. ______________________ Fill in new values in the table for this reaction.

# Atoms in Reactants	Element	# Atoms in Products
	K	
	Cl	
	O	

80. The equation in now balanced for __________ but is not balanced for __________ and __________. This can be corrected by adding a coefficient of __________ in front of __________.

81. Rewrite the balanced equation. ______________________. Cross out the old values in the table above and enter the new values. When an equation is fully balanced, all the numbers on the left are ________________ to the corresponding values on the right.

**Advanced Difficulty

The best way to develop skill in balancing equations is to practice. More examples are given below that can be used for this purpose. The use of tables is a helpful device that promotes a systematic approach to more complex problems.

Questions

82. Balance this equation using the table method. $H_2SO_4 + NaOH \rightarrow Na_2SO_4 + H_2O$

# Atoms in Reactants	Element	# Atoms in Products
	S	
	Na	
	H	
	O	

The table does not need to be rewritten. As each coefficient is added, old values can be crossed out and new ones entered. The answer section shows how such a table would look after the problem has been completed.

83. For this and all additional questions, balance the equation using any method you wish.

$$Mg + O_2 \rightarrow MgO$$

84. $Na + F_2 \rightarrow NaF$

85. $HCl + Ca(OH)_2 \rightarrow CaCl_2 + H_2O$
When a group in brackets has a subscript after it, the subscript refers to the entire group that is in the brackets. In this problem, it means there are two copies of the OH group in the second compound that is listed.

86. $C_6H_{12}O_6 + O_2 \rightarrow CO_2 + H_2O$

87. $Al + O_2 \rightarrow Al_2O_3$

88. Explain why all balancing is done with coefficients and why subscripts in the reactants or products cannot be changed.

Covalent Bonding: An Overview

Our survey of the elements is now complete. You have learned to use a guideline, the octet rule, to predict which elements are stable and which are not. Elements that have an octet of electrons in the valence shell are stable. Elements that do not begin with this configuration will create it during chemical reactions. During these reactions, they either transfer or share electrons to form their valence octet. In this chapter, we have focused on elements that share electrons to achieve stability. Elements with four or five electrons in the valence shell always share electrons to create a valence octet. Elements with six or seven valence electrons share electrons to form an octet when their bonding partner is not an electron donor.

A covalent bond forms when two atoms share a pair of electrons. Covalent bonds join a set of atoms together into a structure called a molecule. Just as one atom is the smallest part of an element that still has the properties of the element, so one molecule is the smallest part of a compound that still has the properties of that compound. Molecules formed as the result of covalent bonding differ in some respects from the compounds formed as a result of ionic bonding. An ion in a crystal lattice interacts on all sides with ions that have an opposite charge. Discrete molecules are not present. One sodium ion is not bonded to just one other chloride ion. In contrast, a set of atoms linked by covalent bonds forms a discrete particle, the molecule. The molecule is a definite unit and is not bonded to other molecules around it in a large lattice.

Molecules have been represented in this chapter by electron dot structures and by structural formulas. Both are useful because they focus on the number of valence electrons and help us determine when the requirements of the octet rule have been met. Neither is a realistic picture of a molecule. Actual molecules are three dimensional, not flat. Furthermore, electrons are not stationary dots or dashes, but are better visualized as clouds. When covalent bonds form, the electron clouds of atoms fuse together and take on new shapes. Space-filling models of molecules represent the shape of the fused electron clouds for the entire molecule. The diagrams below are examples.

Let us again move down to the ultramicroscopic world. As we approach methane, we see roughly spherical electron clouds with four swollen areas on each one. The diagram below represents two views of this shape. Many such molecules are floating about, sometimes bumping into one another and then moving away in a new direction. Let us increase our magnification and move inside one methane molecule. Here we find vast reaches of empty space, just as we did in the single atom. If we magnify enough, we can perhaps find the five tiny specks, the nuclei of the atoms in methane; one would be a carbon nucleus and four would be hydrogen nuclei. Many of you heat your homes with natural gas. As you enjoy the warmth in winter, remember these four-sided clouds with their tiny nuclei inside. The energy stored in these structures is responsible for your comfort.

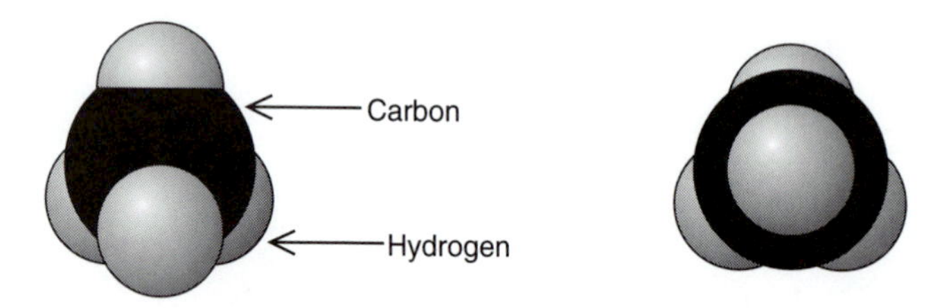

Two Views of Methane: Space Filling Model

Answers

1.

Element	# Valence Electrons	Stable or Unstable	Type of Chemical Reaction Predicted by the Octet Rule
Oxygen	6	Unstable	May accept two electrons to complete valence octet, or shares two electrons
Chlorine	7	Unstable	May accept one electron to complete valence octet, or shares one electron
Carbon	4	Unstable	Shares electrons to complete an octet
Neon	8	Stable	None—this element is stable
Nitrogen	5	Unstable	Shares electrons to complete an octet
Hydrogen	1	Unstable	May lose the valence electron and form a cation (positive), or shares electrons
Calcium	2	Unstable	Loses two valence electrons and forms a cation

2. four or five electrons (share); six or seven electrons (accept electrons or share).

3. gaining (one electron); donate (electrons); one to three (valence electrons).

4. donate; share; four or five (valence electrons); six or seven (valence electrons); four to seven (valence electrons).

5. sharing; one; single.

6. six (valence electrons); gaining; donor. Some examples of electron donors are Na, K, Al, and Ca. Any element that has one, two, or three valence electrons would be a correct answer.

7. No. The two oxygen atoms will share electrons.

   Oxygen atoms will ionize by accepting two electrons, but only if an electron donor is available. No electron donor is available when oxygen atoms react with each other.

8. An element with six or seven valence electrons will form ionic bonds when its bonding partner has one, two, or three valence electrons and can function as an electron donor. In other words, it will form negative ions (anions) when its bonding partner can form positive ions (cations). An element with six or seven valence electrons will share electrons when its bonding partner has four to seven electrons and is not an electron donor. In other words, it shares electrons when its bonding partner is an element that does not normally form positive ions.

9. five (valence electrons); share; three (shared pairs); triple (bond).

10. double (bond); ionic; single (covalent bond); triple (bond).

11. unstable; sharing, gaining, or losing (electrons).

12. 1 (atomic number of hydrogen); one (valence electron); one (dot).

13. 6 (atomic number of carbon); four (valence electrons); C (chemical symbol for carbon); four (dots).

14. five (valence electrons); five (dots).

15. An element is represented by its chemical symbol. Dots are placed around the symbol. The number of dots is equal to the number of valence electrons. Electrons present in completed inner shells are not shown.

16.

Element	# Valence Electrons	Electron Dot Structure
Oxygen	6	$:\underset{\cdot}{\overset{\cdot\cdot}{\text{O}}}\cdot$
Neon	8	$:\underset{\cdot\cdot}{\overset{\cdot\cdot}{\text{Ne}}}:$
Fluorine	7	$:\underset{\cdot\cdot}{\overset{\cdot\cdot}{\text{F}}}\cdot$

The placement of the dots that represent the valence electrons is not critical. When there are four or fewer electrons, the electrons are single. As more electrons are needed, each forms a pair with one of the first four electrons. The position of single electrons versus pairs is not significant. For example, the following four ways of showing fluorine are all equivalent, because the actual atom is spherical and can rotate. The flat projections we write are not accurate pictures of atomic shape.

$:\underset{\cdot\cdot}{\overset{\cdot}{\text{F}}}:$ $\quad :\underset{\cdot\cdot}{\overset{\cdot\cdot}{\text{F}}}\cdot$ $\quad :\underset{\cdot}{\overset{\cdot\cdot}{\text{F}}}:$ $\quad \cdot\underset{\cdot\cdot}{\overset{\cdot\cdot}{\text{F}}}:$

17. five (separate atoms); molecule; five (atoms); covalent bonds.

18. molecule.

19. four (electrons); one (electron); eight (electrons total); eight (electrons in the structure).

20. Eight (electrons). Yes, the structure should be stable.

    The octet rule states that elements tend to be stable when they have eight valence electrons. Carbon does not have eight electrons on its own. After forming four single covalent bonds, it now shares in eight electrons (two electrons are shared with each of four hydrogens) and, therefore, is stable.

21. Two (electrons). Yes, this structure should be stable.

    Hydrogen is a small atom that uses only the first electron shell. This shell is filled when it contains two electrons. In the case of those elements that use only the first electron shell, when the first electron shell is filled, the atom or ion is stable.

22. single covalent bond. Yes.

    Electrons that are between two atoms contribute to the electron cloud of both atoms simultaneously. Shared electrons belong to both the participating atoms. Shared electrons are included when determining whether the octet rule is satisfied for both the participating atoms.

23. one (electron); one (complete inner shell); two (electrons); no

24. four (single covalent bonds); one (shared pair); one (dash).

25. double (covalent bond); two (dashes).

26. Eight (atoms); four (valence electrons); one (valence electrons); 14 electrons.

    (Each carbon contributes four electrons, and the six hydrogens each contribute one.)

27. Write electron dot structures for all the carbons and hydrogens that are components of the ethane structure. Write them as separated atoms.

    ·Ċ· ·Ċ· H· H· H· H· H· H·

28. octet (of electrons); a pair of (electrons).

29.

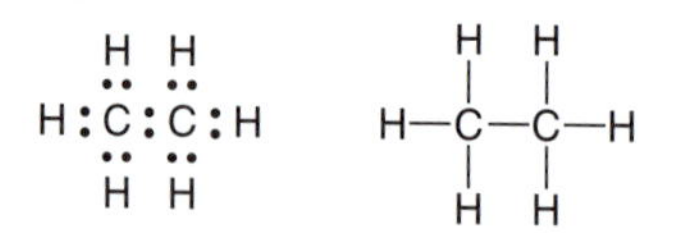

30. Only valence electrons participate in forming covalent bonds.

    Although inner-shell electrons may be present in some of the participating atoms, it is simpler not to show them, because they are not involved in the bonding process.

31.

Compound	Chemical Formula	Electron Dot Structure of Atoms	Electron Dot Structure of Molecule	Structural Formula
Water	H_2O	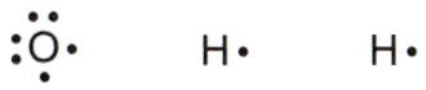 :Ö· H· H·	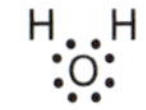 H:Ö:H	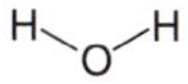 H–O–H
Ammonia	NH_3	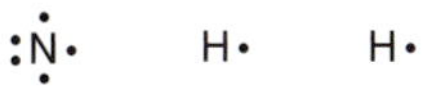 :Ṅ· H· H·	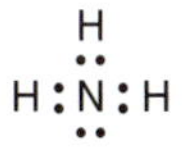 H:N̈:H with H above	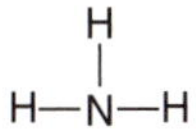 H–N–H with H above
Oxygen	O_2	:Ö· :Ö·	 :O::O:	O=O
Carbon Dioxide	CO_2	 ·Ċ· :Ö· :Ö·	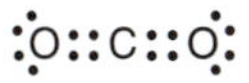 :O::C::O:	O=C=O

Note: Structural formulas show each covalent bond with a dash. One dash represents a single covalent bond (one shared pair of electrons). Valence electrons that are nonbonding are not shown.

32. Eight (electrons around oxygen); eight (electrons around carbon); eight (electrons around nitrogen); two (electrons near each hydrogen).

33. Atoms that do not have an octet of electrons on their own may acquire an octet by sharing electrons.

    Most elements share enough electrons to create an octet in the valence shell. Hydrogen is an exception. It is a small element that normally has electrons only in the first electron shell. This shell is full when it contains two electrons, and hydrogen is stable when this first shell is filled.

34. Hydrogen has: one (valence electron); one (shared electron); one (covalent bond).

35. Oxygen has: six (valence electrons); two (shared electrons); two (covalent bonds).

    (This requirement may be satisfied with two single covalent bonds or one double bond.)

36. Nitrogen has: five (valence electrons); three (shared electrons); three (bonds formed).

    (Nitrogen forms three bonds. This can be accomplished with three single covalent bonds, one triple bond, or one double plus one single covalent bond.)

37. Carbon has: four (valence electrons); four (shared electrons); four (covalent bonds).

(Carbon can meet its bonding requirements with any combination of single, double, and triple bonds, as long as the total is four bonds.)

38.

	H	O	N	C
# Valence Electrons	1	6	5	4
# of Electrons Needed for Stability	1	2	3	4
# Covalent Bonds Formed	1	2	3	4

Note that when the four elements are listed in order, H O N C, they spell a word (if you don't mind unconventional spellings). The number of covalent bonds formed is then 1 - 2 - 3 - 4. Organic molecules such as carbohydrates, proteins, and lipids are formed principally of these four elements. Some of these molecules have large and complex structures, yet the individual atoms of which they are composed still follow these guidelines.

39. Nonpolar. The two atoms are the same, so they have equal tendencies to pull electrons. The electrons are therefore shared equally, forming a nonpolar covalent bond.

40. different; unequally.

41. (toward) oxygen; (away from) carbon.

42: negative; negatively.

43. negatively; negative; oxygen (is slightly negative).

44. away; positive.

Because the negative electron cloud is slightly displaced away from the carbon, it creates a slightly positive charge on the carbon end of the structure. The carbon atom has a nucleus with positively charged protons in it, which are not fully shielded and neutralized by the electron cloud after the cloud has shifted away.

45. toward (oxygen); away (from carbon); positive (pole near carbon); negative (pole near oxygen; polar (bond).

(The use of the terms "pole" and "polar" is analogous to using the same terms to refer to the north and south magnetic poles of the earth, or to the two poles of a magnet.)

46. 3.0 (electronegativity of nitrogen); 2.1 (electronegativity of hydrogen); polar (bond); nitrogen (direction of electron cloud shift).

47.

A Bond Between Atom 1 and Atom 1	Type	Location of Electron Cloud
(1) Carbon Bonded to (2) Carbon	Nonpolar	Shared equally
(1) Carbon Bonded to (2) Hydrogen	Polar	Shifted slightly toward carbon
(1) Carbon Bonded to (2) Oxygen	Polar	Shifted toward oxygen
(1) Oxygen Bonded to (2) Oxygen	Nonpolar	Shared equally
(1) Oxygen Bonded to (2) Nitrogen	Polar	Shifted slightly toward oxygen
(1) Sulfur Bonded to (2) Carbon	Nonpolar	Shared almost equally
(1) Nitrogen Bonded to (2) Nitrogen	Nonpolar	Shared equally

Explanation: Any bond between two atoms of the same type is nonpolar. On average, the electron cloud is located midway between the two sharing atoms. If two atoms differ only slightly in electronegativity, the bond between them will be mainly nonpolar as well. The bond between sulfur and carbon is an example. As the difference in electronegativities increases, the bond becomes more polar. Bonds between hydrogen and carbon are examples of slightly polar bonds. If the difference in electronegativities is great, the bond is strongly polar. The bond between carbon and oxygen is an example. This means that the electron cloud shifts farther toward the electronegative atom and is closer to the ionic situation.

48. Elements with high electronegativities are located in the upper right in the periodic table.

49. Elements that need only one or two electrons to complete an octet in the valence shell tend to have a high electronegativity. Elements that are small also tend to have a high electronegativity.

In small elements, there are fewer inner electron shells that are filled. Filled inner shells tend to shield the positive nucleus. Without many filled shells, the pull of the positive nucleus leads to greater electronegativity in the smaller atoms.

50. 2.1 (electronegativity of H); nonpolar; midway between the hydrogens.

51. 2.1 (electronegativity of H); 2.5 (electronegativity of C); 0.4 (difference between the two values); slightly polar.

52. The carbon end is indicated as the partially negative pole. The hydrogen end is indicated as the partially positive pole. The δ^- symbol designates the negative end of the structure, and the δ^+ symbol indicates the positive end of the structure.

    Because carbon is more electronegative, the electron cloud shifts toward it, which explains why the carbon is the negative pole. Since the electrons shift away from hydrogen to a slight extent, it becomes the positive pole. These are partial charges.

53. 3.5 (electronegativity of O); 2.1 (electronegativity of H); 1.4 (difference between the two values); strongly polar.

54. oxygen (negative pole); hydrogen (positive pole); neutral overall; oxygen; polar.

55. δ^- (negative pole); δ^+ (positive pole); partial.

56. carbon and hydrogen (reactants); methane (product); arrow; no.

    Chemical reactions occur at many different rates. Some are so fast that they are explosive. Some occur at such slow rates that reaction is barely detectable. All reactions are represented by an arrow, whether they are fast or slow. Other information is necessary to determine the speed of the reaction.

57. $Na + Cl \rightarrow NaCl$; sodium (Na) and chlorine (Cl) are reactants; product is sodium chloride (NaCl).

58. calcium (Ca) and chlorine (Cl_2—the diatomic molecule) are reactants; calcium chloride ($CaCl_2$) is the product; no. Cl_2 is an example of a diatomic molecule. Two atoms of chlorine share electrons and are linked by a covalent bond. Chlorine molecules are fairly stable, whereas atoms of chlorine are highly reactive.

59. products (on right); reactants (on left).

60. methane and oxygen (reactants); CH_4; subscript. The subscript 4 indicates that there are four atoms of hydrogen bonded into the methane molecule. One molecule of methane is used.

    (There is no coefficient in front of this molecule. When there is no coefficient, a 1 is understood.)

61. subscript (smaller 2); atoms (in one molecule); diatomic molecule; coefficient; 2 (oxygen molecules); 4 (oxygen atoms). (2 atoms per molecule times 2 molecules, or $2 \times 2 = 4$.)

62. carbon dioxide and water (products for the reaction); one (molecule of CO_2); two (molecules of H_2O).

63. One (carbon in the reactants); one (carbon in the products); four (hydrogens in the reactants); four (hydrogens in the products); four (oxygens in the reactants); four (oxygens in the products).

    (Note that the two oxygen atoms from CO_2 and the two oxygen atoms from H_2O must be added to obtain the total on the product side of the equation.)

64. All the atoms that are present in the reactants are present in the products. No atoms are lost or gained during a chemical reaction; they are only rearranged, and the bonding pattern changes. Therefore, the number of each kind of atom is the same before and after the chemical reaction.

65. H_2 (formula for hydrogen); O_2 (formula for oxygen); H_2O (formula for water); $H_2 + O_2 \rightarrow H_2O$.

66. Two (oxygen atoms present in the reactants); one (oxygen atoms present in the product); two (molecules of water needed); $H_2 + O_2 \rightarrow 2H_2O$.

    (The coefficient 2 indicates that there are two water molecules on the product side. The coefficient applies to the entire molecule that follows it, not just the first atom.)

67. Two (hydrogens on the reactant side); four (hydrogens on the product side); (not) balanced; two (hydrogen molecules required); coefficient.

68. $2H_2 + O_2 \rightarrow 2H_2O$. This equation states that two molecules of hydrogen and one molecule of oxygen react to produce two molecules of water.

69. Yes. A balanced equation has the same number and kinds of atoms on the left side of the equation (in the reactants) and on the right side of the equation (in the products).

70. Oxygen bonds to other oxygen atoms (in O_2 in the reactants). Oxygen bonds to hydrogen (in the products); hydrogen is bonded to other hydrogen atoms (in H_2 the reactants). Hydrogen is bonded to oxygen (in the products).

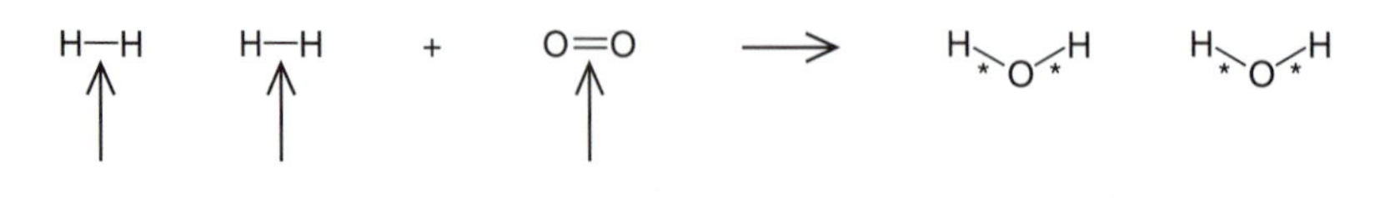

71. $CO_2 + H_2O \rightarrow C_6H_{12}O_6 + O_2$

72. One (carbons are present on the reactant side); six (carbons on the product side); no (not balanced); six (coefficient required); in front of CO_2

73. $6CO_2 + H_2O \rightarrow C_6H_{12}O_6 + O_2$; (There are now six carbons on each side of the equation.) Two (hydrogens in the reactants); 12 (hydrogens in the products). No (not balanced for hydrogen).

74. $6CO_2 + 6H_2O \rightarrow C_6H_{12}O_6 + O_2$; Because there are twelve hydrogens in the products, it is necessary to have twelve hydrogens in the reactants. Each water contains two hydrogens bonded in its structure. Six water molecules thus contribute the required 12 atoms of hydrogen. ($2 \times 6 = 12$).

75. 18 (oxygens in the reactant molecules); 8 (in the product molecules). No, (not balanced for oxygen).

The number of oxygens in the reactants is calculated as follows: There are two O atoms per CO_2 and a total of six CO_2 molecules. The total O atoms from CO_2 is therefore 12. There is one oxygen per water molecule and a total of six water molecules. Therefore there are six O atoms contributed by water. The total for O is thus 12 (from CO_2) + 6 (from water) for a total of 18.

76. $6CO_2 + 6H_2O \rightarrow C_6H_{12}O_6 + 6O_2$; Six molecules of carbon dioxide and six molecules of water react. One molecule of sugar (glucose) and six molecules of oxygen are the products.

77. $KClO_3 \rightarrow KCl + O_2$.

# Atoms in Reactants	Element	# Atoms in Products
1	K	1
1	Cl	1
3	O	2

78. Oxygen (not balanced in the above equation); 3 (atoms in the reactants, on the left); 2 (atoms in the products, on the right);

Multiplying the values of 2 and 3 gives a lowest common multiple of 6. The way to balance this equation for the element of oxygen is to have six atoms of oxygen on each side. Because $KClO_3$ has three atoms of oxygen, two molecules of this compound will be needed. A coefficient of 2 is therefore required. To get six atoms of oxygen on the product side, a coefficient of 3 is needed in front of O_2 (3 molecules $\times$ 2 atoms per molecule = 6 atoms).

79. Equation with new coefficients is: $2KClO_3 \rightarrow KCl + 3O_2$
New values in the table for this reaction are:

# Atoms in Reactants	Element	# Atoms in Products
2	K	1*
2	Cl	1*
6	O	6

80. oxygen (is balanced); potassium and chlorine (not balanced); 2 (coefficient); KCl.

81. $2KClO_3 \rightarrow 2KCl + 3O_2$ The two values of 1 on the right are indicated by an asterisk. These should be crossed out and replaced with a 2. equal (values on left and right).

82. Balance this equation using the table method. The balanced equations is:

$$H_2SO_4 + 2NaOH \rightarrow Na_2SO_4 + 2H_2O$$

# Atoms in Reactants	Element	# Atoms in Products
1	S	1
~~1~~ 2	Na	2
~~3~~ 4	H	~~2~~ 4
~~5~~ 6	O	~~5~~ 6

The table is filled in first with the values for the unbalanced equation. As coefficients are added, old values are crossed out and new corrected ones are added.

83. $Mg + O_2 \rightarrow MgO$ $2Mg + O_2 \rightarrow 2MgO$

84. $Na + F_2 \rightarrow NaF$ $2Na + F_2 \rightarrow 2NaF$

85. $HCl + Ca(OH)_2 \rightarrow CaCl_2 + H_2O$ $2HCl + Ca(OH)_2 \rightarrow CaCl_2 + 2H_2O$

86. $C_6H_{12}O_6 + O_2 \rightarrow CO_2 + H_2O$ $C_6H_{12}O_6 + 6\ O_2 \rightarrow 6CO_2 + 6H_2O$

87. $Al + O_2 \rightarrow Al_2O_3$ $4Al + 3O_2 \rightarrow 2Al_2O_3$

88. Coefficients indicate the numbers of each type of compound or atom that are used and produced in chemical reactions. Any proportions necessary to achieve balance are possible. The proportions present in a particular chemical reaction are determined by the conservation of matter. Subscripts cannot be changed to achieve balance. Changing a subscript would change the identity of the chemical compound. The identity of the chemical compound, and hence the subscripts, is determined in other ways that have already been covered.

Chapter Test: Covalent Bonding

The questions in this chapter test evaluate your mastery of all the objectives for this unit. Material from earlier chapters is not tested directly, but recall of material presented previously is necessary background in some instances. Take this test without looking up material in the chapter. You may use the periodic table. Questions that test intermediate or advanced objectives are indicated by asterisks, and you may omit them if you did not cover those topics in the chapter. After you have completed the test, you can check your work by using the answer key located at the end of the test.

1. Elements with four or five valence electrons tend to achieve stability by ______________________.

2. Elements with six or seven valence electrons may ionize by ___________ electrons when they react with elements that will ___________ electrons. Elements with six or seven valence electrons will achieve stability by sharing electrons when they react with elements that have ___________ valence electrons.

3. One shared pair of electrons is a ___________. Six shared electrons form a ___________.

4. Explain why atoms sometimes transfer or share electrons. ______________________________

   ____

5. Write electron dot structures for the elements listed below:

   Argon

   Bromine

6. Propane has the formula C_3H_8. In the space below, write electron dot structures for the atoms that are components of this compound. Then write an electron dot structure for the compound. Finally, write the structural formula for this compound.

7. In the space below, write the chemical formula for water. Then write the electron dot structure and the structural formula for water.

   What are the major differences among these three ways of representing the water molecule?

   ____

   ____

8. Nitrogen typically forms ______ covalent bonds. Carbon forms ______ covalent bonds. In each case, there is an ______ of valence electrons around each atom after the bonds have formed.

9.*Define the term "electronegativity." ____

   ____

10.*When two atoms have exactly the same electronegativity, the bond formed between them will be ___________.

11.*The electronegativity of N is 3.0, and that of H is 2.1. A bond between these two elements is ___________, and the electrons have shifted toward ___________.

12.*The Greek letter δ means ___________. Explain how this symbol is used to designate the poles of certain chemical bonds.

   ____

13. The chemical formula NaCl tells us that sodium and ___________ are present in the compound ___________, in the proportions ___________. The chemical formula $C_2H_5O_2N$ tells us that the elements ___________ are present in the compound glycine, in the proportions ___________. The formula $C_2H_5O_2N$ also tells us that a molecule of glycine contains ___________ atoms, all linked by covalent bonds into ___________.

14. Write an equation that represents the reaction of hydrogen and oxygen to produce water. Then balance the equation.

____

15. The following questions refer to this formula: 7 $C_6H_{12}O_6$

    Which number(s) are subscripts? ____________

    What do the subscripts mean? ____

    Which of the number(s) are coefficients? ____________

    What do coefficients mean? ____

16. When the number and kinds of atoms are the same on both sides of an equation, the equation is said to be ____________.

17.* Sugar and oxygen can be metabolized within many cells. Carbon dioxide and water are the products. This reaction sequence is called cellular respiration. Write an equation to represent this process.

____

Balance this equation. Show the individual steps.

18. State the octet rule. ____

____

19. Predict the behavior of elements that do not have an octet of valence electrons.

____

____

20.**Propane, C_3H_8, and O_2 can react to produce CO_2 and H_2O. Write a balanced equation to represent this reaction.

21. During the course of a chemical reaction, bonds are ____________ in the reactants and then __________________ to create the products.

Answers

1. Sharing enough electrons to create an octet in the valence shell.

2. gaining (electrons); donate (elements with one, two, or three valence electrons can function as electron donors); four to seven (number of valence electrons in bonding partners that lead to sharing electrons).

3. Single covalent bond (one shared pair); triple bond (three shared pairs).

4. Atoms that do not have an octet of electrons in the valence shell are unstable, achieving stability by transferring or sharing the number of electrons required to create an octet.

5.

:Ar: Argon   :Br· Bromine

6. Propane

·C· ·C· ·C· H· H· H· H· H· H· H· H·

H:C:C:C:H   H—C—C—C—H

7. Water: chemical formula is H_2O.

H :O: H   H—O—H

Structural formulas show which elements are present and how many atoms of each type. In electron dot structures, all valence electrons are shown, whether or not they are involved in chemical bonding. In structural formulas, covalent bonds are shown with dashes. Nonbonding valence electrons are not shown.

8. Three (bonds in nitrogen); four (bonds in carbon); octet (number of valence electrons after bonding).

9.*Electronegativity is a measure of the tendency of an atom to attract electrons. The higher the number, the more strongly the atom pulls on electrons.

10.*nonpolar (covalent).

11.*polar covalent (bond between N and H); nitrogen (electrons shift to it).

12.*partial. The symbol δ^+ is used to designate the slightly positive pole of a polar covalent bond. The symbol δ^- is used to designate the slightly negative pole of a polar covalent bond.

Although a polar covalent bond is neutral overall, the poles form because the electron cloud is shifted toward the more electronegative element, creating a slight internal separation of charge.

13. Chloride; sodium chloride; 1:1; carbon, hydrogen, oxygen, and nitrogen; 2: 5: 2: 1. 10 (atoms); 1 molecule.

The formula for a compound containing ionic bonds (such as NaCl) states the proportions of the elements that are present in the compound. Molecules are not defined for such compounds. As you learned in the previous chapter, an ion is surrounded on all sides by ions with an opposite charge. Attractions between ions of opposite charge are the ionic bonds. One ion, however, is not attached to only one other ion; rather, it is part of a crystal lattice with ionic interactions on all sides. In contrast, molecules are present in compounds that rely on covalent bonding (glycine is an example. A specific group of atoms share electrons, form a discrete particle called a molecule, and function as a unit.

14. $H_2 + O_2 \rightarrow H_2O$ Equation $2H_2 + O_2 \rightarrow 2H_2O$ Balanced equation

15. 7 $C_6H_{12}O_6$

The smaller numbers (6, 12, and 6) are subscripts. The subscripts tell how many atoms of a particular type are present. For example, the small 6 after C indicates that the molecule contains 6 atoms of carbon bonded into the structure of the molecule. The number 7 is a coefficient. The coefficient tells how many molecules are present. It refers to the entire structure that follows.

The 7 means that there are 7 molecules of sugar, each of which contains 6 atoms of carbon, 12 atoms of hydrogen, and 6 atoms of oxygen.

16. balanced.

17.*Cellular respiration equation: $C_6 H_{12} O_6 + O_2 \rightarrow CO_2 + H_2O$

Balancing the equation: The first step is to balance carbon. Because there are six carbons on the reactant side of the equation, there must be six on the product side. This requires a coefficient of 6 in front of the carbon dioxide, as follows: $C_6 H_{12} O_6 + O_2 \rightarrow 6CO_2 + H_2O$

The next step is to balance hydrogen. This requires a coefficient of 6 in front of water, and the equation now becomes: $C_6 H_{12} O_6 + O_2 \rightarrow 6CO_2 + 6H_2O$

The equation is now balanced for carbon and hydrogen, but not for oxygen. There are eighteen oxygens on the product side and only eight oxygens on the reactant side of the equation. To balance this, a coefficient of 6 is required in front of oxygen. The balanced equation now becomes: $C_6 H_{12} O_6 + 6O_2 \rightarrow 6CO_2 + 6H_2O$

18. The octet rule states that an element having eight valence electrons is stable. Elements that do not have an octet of valence electrons are not stable.

19. Elements with one to three valence electrons will lose those electrons, exposing an octet in the electron shell below. Elements with four or five valence electrons will share electrons to form a valence octet. Elements with six or seven valence electrons will accept electrons, completing an octet when an electron donor is available. When a donor is not available, these elements share electrons to create a valence octet.

20.**$C_3 H_8 + 5O_2 \rightarrow 3CO_2 + 4H_2O$

21. broken; formed.

Chapter 4

Aqueous Chemistry

Objectives

1. Define the following terms and apply the definitions correctly: acids, aqueous, bases, bond angle,* buffers, colloidal dispersion, electrolyte, hydrogen bond, molarity, molecular weight, mole, neutralization, nonpolar molecule, pH, polar molecule, salts, solute, solution, solvent, and tetrahedral shape.*
2. *Predict which molecules will be polar and which will be nonpolar.
3. Sketch a water molecule, correctly showing the shape and the position of partial charges.
4. Predict which molecules can form hydrogen bonds.
5. List the unique properties of water. Explain how these properties result from the ability of water to form hydrogen bonds.
6. Predict which molecules will be soluble in water.
7. Determine which solutes are electrolytes. Classify each electrolyte as an acid, base, or salt.
8. Use the pH scale to describe the acid or base strength of a solution.
9. Predict which reactants will undergo a neutralization reaction. Predict the products that will form.
10. Write equations to describe neutralization reactions.
11. *Balance equations for neutralization reactions.
12. *Explain how a buffer system works.
13. Calculate the molecular weight of a compound.
14. **Calculate and explain how to prepare solutions of a specified molarity.
15. Solve dilution problems.

The Importance of Water

Water is one of the most abundant compounds on the earth. It is present in oceans, other bodies of water, and as water vapor in the atmosphere. Water is an excellent solvent, a compound that can dissolve other materials. Many kinds of atoms, molecules, and ions can dissolve in water, forming an aqueous solution. Materials that dissolve in a solvent are called solutes. Seawater is an example of a complex aqueous solution that contains several dissolved salts. Many living organisms live in water, and all life-forms require water.

Some large molecules, such as proteins, form colloidal dispersions in water. The proteins do not become fully dissolved in water, but are stably suspended in it, and do not settle out. Colloidal dispersions are often translucent in appearance and gelatinous in texture. Cytoplasm, the material that fills living cells, contains many dissolved and dispersed substances. Cytoplasm is therefore both a complex solution and a colloidal dispersion. An understanding of cells and their biochemistry thus requires that we thoroughly understand the chemistry of water.

Questions

1. Aqueous is a term that refers to ____________.

2. An aqueous solution is a mixture of materials. The most abundant compound is the solvent ____________ that dissolves the other materials, called ____________. Water is an excellent ____________, because many other materials dissolve in it and form ____________.

3. Both common table salt and sugar dissolve readily in water. The solute(s) would be ____________, and the solvent(s) would be ____________. Describe the appearance of these solutions.

4. Give two examples of complex aqueous solutions and explain why they are important.

5. Certain large molecules such as ____________ become stably dispersed in water. This mixture is called a __________________, and it may be gelatinous in texture and ____________ in appearance.

6. Give an example of a colloidal dispersion.

7. There are more ____________ molecules in living cells than any other types of molecules.

Molecular Shape

The shape and polarity of the water molecule determine many of the chemical properties of water. It is easiest to understand the shape of the water molecule by first considering a simple organic compound, methane. The methane molecule, CH_4, is often represented with a structural formula showing it as a flat or planar molecule. Methane molecules actually are three-dimensional, having the shape of a regular tetrahedron as represented below.

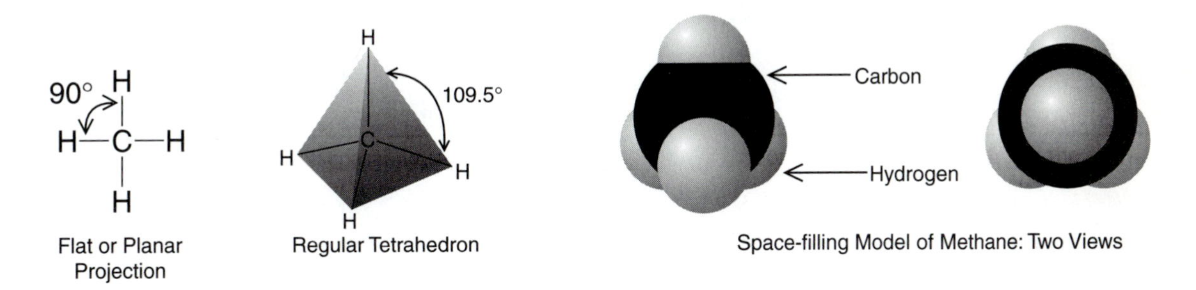

Flat or Planar Projection

Regular Tetrahedron

Space-filling Model of Methane: Two Views

Questions

*Some questions in this section are intermediate or advanced in difficulty.

8. How many surfaces, or faces, does the regular tetrahedron have? ____________

9.* What is the shape of one of the tetrahedral surfaces, and how do the different faces compare to each other?

10. What atom is at the center of the methane molecule? ____________

    Where is this atom located in the tetrahedral model of methane? __________________

    What atoms are bonded to the central atom, and where are they located in the tetrahedral model?

11. What connects the H and C atoms in the methane molecule? ________________________________

    How are these represented in the flat projection formula? ________________________________

12. What is a covalent bond? ____________. The charge on ____________ is negative. Charges with the same sign ____________ each other. Charges that are ____________, push as far away from each other as possible.

13.**Given the fact that electrons repel each other, explain why the methane molecule assumes the shape of a regular tetrahedron.

Ammonia and water molecules exhibit tetrahedral symmetry, just as methane does. Nitrogen and oxygen can complete an octet of electrons in their valence shells by forming covalent bonds with hydrogen. The four pairs of electrons in the valence shell then occupy the four positions at the points or apexes of a regular tetrahedron. Electron pairs behave this way whether or not they are participants in a covalent bond.

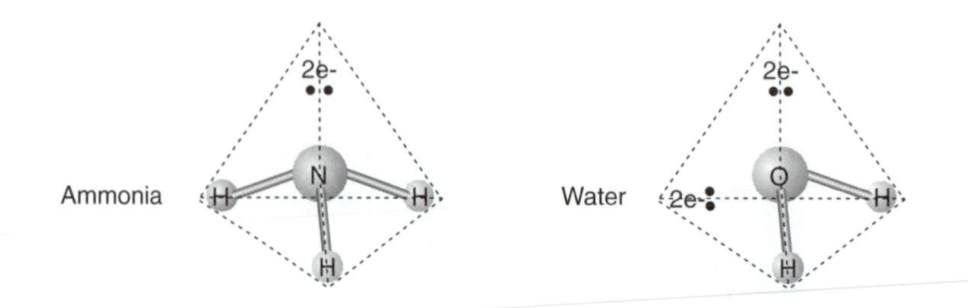

Questions

*Some questions within this section are intermediate or advanced in difficulty.

14. An atom of ____________ is at the center of the ammonia molecule. An atom of ____________ is at the center of the water molecule. These atoms are located __________________ of the tetrahedron.

15. Three atoms of ____________ are bonded to the central atom in ammonia. Two atoms of ____________ are bonded to the central atom in water. Where are these atoms located on the tetrahedron?

16. How many pairs of unbonded electrons are present in the valence shell of nitrogen (in the ammonia molecule)? ____________ There are ____________ pair(s) of unbonded valence electrons in the water molecule. Where are the unbonded electrons located on the tetrahedron?

17.*Explain why ammonia and water assume a tetrahedral shape.

18.*Inspect the flat projection diagram of methane (page 143). The angle between two of the hydrogens is called the bond angle. If methane were really a flat molecule, the bond angle would be ____________.

19.*Methane is not a flat molecule; rather, it has a __________________ shape. The bond angle in methane is therefore ____________. The tetrahedral angle of ____________ is greater than the square or flat angle of ____________. This permits the pairs of electrons, which are __________________, to get __________________.

20.**One might imagine that the formula for water could be written as H - O - H. Explain why water is not a flat or linear molecule and why this is not an accurate way to represent its structure.

Polar and Nonpolar Molecules

Methane is an example of a nonpolar molecule. Like all molecules, methane has charged particles within it. As a whole, the molecule is neutral, because the number of protons and the number of electrons is equal. Within the molecule, there is no significant separation of charge. All the negative charge (due to the electron cloud) centers at the same point as all the positive charge (due to the protons). There is not a negative pole that is separate from a positive pole. Other hydrocarbons (molecules containing only hydrogen and carbon) generally share this property of nonpolarity.

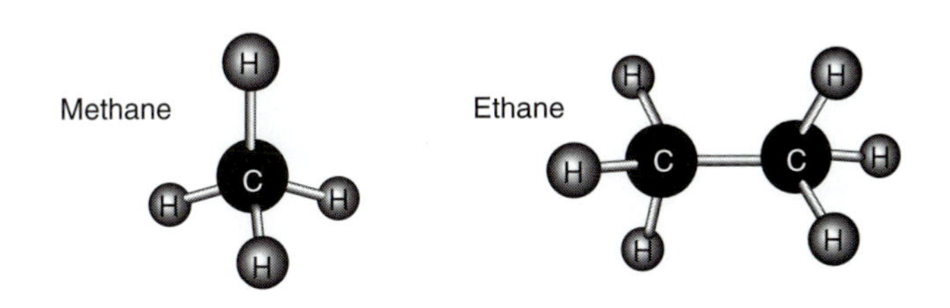

Questions

21. What kind of bond links hydrogen to carbon in the methane molecule? ____________________

22.* Are the carbon-to-hydrogen bonds polar or nonpolar? How can you tell?

23.* Electrons are slightly displaced toward the __________ and away from __________ in the methane molecule.

24.* All the bond angles in the methane molecule are __________. The bond angles in methane __________ the same. All four bonding positions in methane are occupied by __________. The molecule therefore is ______ is not ______ symmetrical.

25.** The bonds within the methane molecule are slightly __________. The center of positive charge in the methane molecule is in the __________. The center of negative charge in the methane molecule is in the __________. When the centers of positive and negative charge are at the same spot, the molecule as a whole is nonpolar. There is not a significant internal separation of charge in the methane molecule.

26.** The methane molecule as a whole is nonpolar, even though the bonds within it are slightly polar. Explain this phenomenon.

27. The ethane molecule is a slightly larger hydrocarbon than methane. Would you predict this molecule to be polar? Explain your answer.

Water is an example of a polar molecule. It has a slightly negative pole and a slightly positive pole, as shown in the diagrams below. Overall, the electron cloud has shifted toward oxygen and away from hydrogen. The shape of the water molecule and the polar covalent bonds within it are responsible for this.

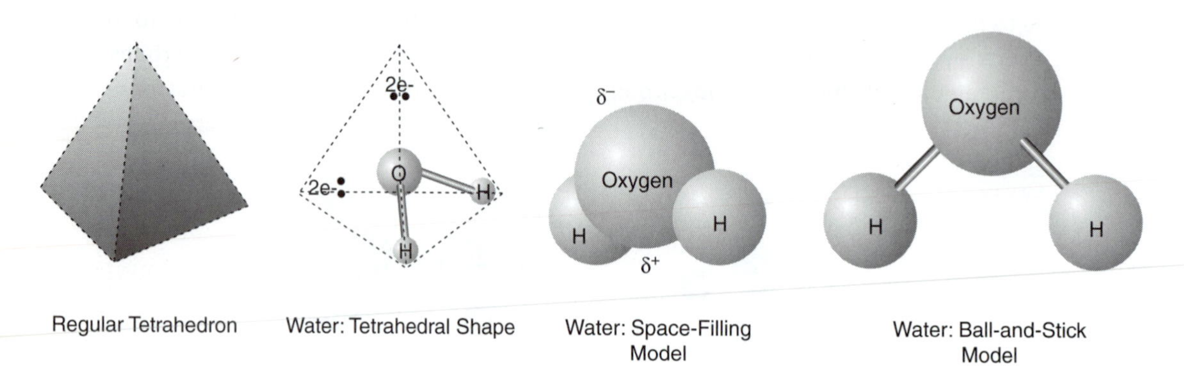

Regular Tetrahedron | Water: Tetrahedral Shape | Water: Space-Filling Model | Water: Ball-and-Stick Model

Questions

28.* The chemical bonds linking H to O in the water molecule are ________________. The electron cloud is shifted toward ____________ with an electronegativity of ____________ and away from ____________ with an electronegativity of ____________.

29. In the water molecule, there are ____________ pairs of valence electrons that are not involved in chemical bond formation. These electron pairs are located at ____________ in the tetrahedron.

30. The two hydrogens in water are located ____________ in the tetrahedron.

31. Describe the overall shape and symmetry of the water molecule.

32. The electron clouds within water are shifted toward ____________ and away from ____________. The center of negative charge in water is near ____________. When the electron cloud shifts toward ____________, it partially exposes some positive charges in the nuclei of the ____________ atoms. The partial positive charge in the water molecule is centered near ____________.

33. What is the definition of a polar molecule? ____

Does water fit this definition? Explain your answer. ________________________________

34. What does the δ^+ near to the hydrogens in water signify? ______________________________

What does the δ^- near the oxygen in water signify? ________________________________

*Intermediate Difficulty

Molecules containing nonpolar bonds are nonpolar. Molecules containing polar covalent bonds may or may not be polar, depending on the geometry and symmetry of the molecule. If the electron clouds shift in opposite directions and offset each other, no poles develop, and the molecules are nonpolar. Carbon dioxide, methane, and other hydrocarbons are examples of this type of molecule. If the geometry of the molecule is such that the displacements of the electron clouds do not cancel each other, poles will develop in the molecule. When an internal separation of charge develops in this way, the molecule is polar. Water and ammonia are examples of this type of molecule.

Questions

35. The carbon dioxide molecule has the structure O = C = O. The bonds that link carbon and oxygen are quite ____________, with the electron cloud shifted toward oxygen. Yet the molecule as a whole is nonpolar. Explain why this is so.

36. In water molecules, the two hydrogen atoms are linked to oxygen by ____________ bonds. The electron cloud is shifted toward ____________ and away from hydrogen. The molecule as a whole is ____________. Explain why this is so, and explain how the situation differs from that of carbon dioxide.

The Properties of Water

Water molecules have a tetrahedral shape (shown in the diagrams on page 129) and are polar. Molecules that are polar have a partial internal separation of charge, forming a pole that is slightly negative and a pole that is slightly positive. The slight opposite charges in different water molecules attract each other. The weak attractions between different water molecules are called hydrogen bonds and are represented by dotted lines in the diagram below.

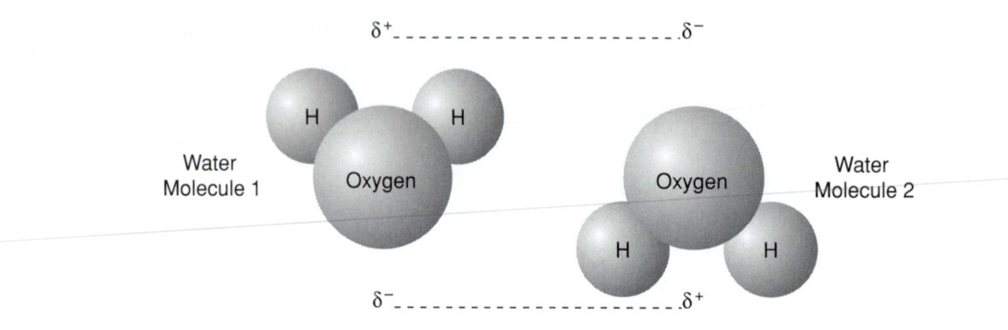

Questions

37. The partially negative pole of water is near the ____________ atom, and the partially positive pole of water is near the ____________ atoms.

38. In the diagram above, there is a weak attraction shown between the oxygen of water molecule 1 and the ____________ of water molecule 2. This weak attraction occurs because opposite charges are ____________ to each other.

39. ____________ bonds form when positive ions and negative ions are attracted to each other. This type of interaction may be quite strong.

40. The attractions shown between different water molecules are represented by ____________. These are ____________ attractions, rather than strong ones.

41. Explain why the attractions between different water molecules are weak interactions rather than strong ones like those seen in ionic bonding.

42. A collection of water molecules are attracted to each other as shown above. These weak attractions are called hydrogen bonds. Do hydrogen bonds link atoms into molecules? Explain your answer.

    Yes ______ No ______

Hydrogen bonds are weak attractions between different molecules, or between distant parts of a large molecule that can fold back on itself. The following conditions are required for hydrogen bond formation: hydrogen must be present in a strongly polar covalent linkage (only bonds to fluorine, oxygen, or nitrogen are sufficiently polar), and the molecule as a whole must be polar or contain large polar regions.

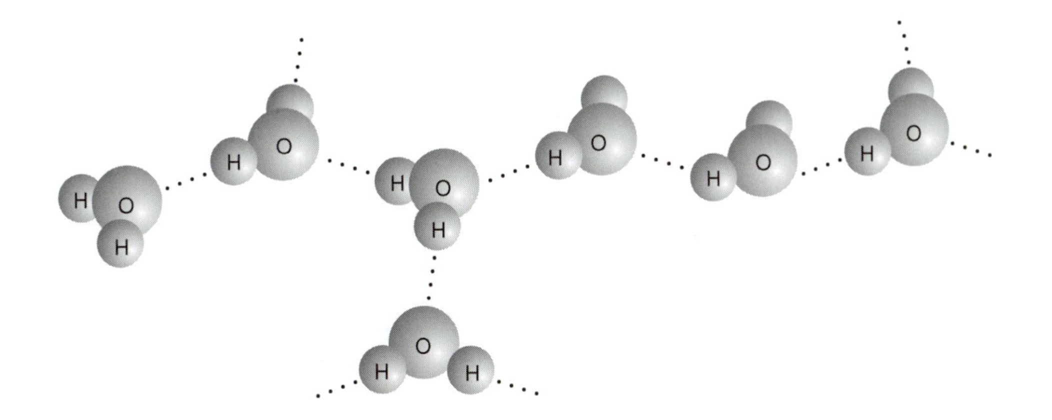

Questions

43. Hydrogen sulfide, H_2S, is a molecule similar to water in some respects. Would you predict that this molecule could form hydrogen bonds? Explain your answer.

    Yes ______ No ______

44. Methane, CH_4, contains hydrogen and has slightly polar covalent bonds. Predict whether it will form hydrogen bonds, and explain the reasons for your answer.

45. Proteins are very large organic molecules (proteins will be covered in Chapter 6) that contain a variety of chemical groups. Amine groups ($-NH_2$) and carbonyl groups ($-C=O$) are among them. Would you predict that hydrogen bonds could form between protein molecules?

    Yes ______ No ______

    Proteins are large and can fold and bend. Given that fact, can you imagine situations that would lead to hydrogen bond formation within a protein molecule? Explain.

46. The following questions refer to the diagram on page 135 (real water molecules do not have such an orderly arrangement). A water molecule is hydrogen bonded to ____________ other water molecules. Actual water, in three dimensions, would hydrogen bond to ____________ (more or less) molecules, because there would be additional layers of water in front of and behind the ones shown in the diagram on page 135.

47. A water molecule is not a completely free and unattached structure, because it forms many ________________. Water is an associated liquid. Explain what this means.

48. Compounds that can form hydrogen bonds have different physical properties from those that cannot. H_2O ____________ form hydrogen bonds and is an ____________ liquid. H_2S ____________ form hydrogen bonds and ____________ an associated liquid. Water and hydrogen sulfide do ________ do not ________ have very similar physical properties.

Hydrogen bonds have a significant impact on the physical properties of water. Compared to other molecules of a similar size, water melts and vaporizes at high temperatures (0° and 100° C respectively). Water has an unusually high heat capacity, as it takes a lot of energy to heat it up. These characteristics are the result of hydrogen bonding.

Questions

49. Temperature is a reflection of the speed of molecular motion. When a substance is hot, its molecules are generally moving faster, whereas at cooler temperatures, molecules, on average, are moving ________.

50. Hydrogen bonds in water are weak ________________________. Molecules of water tend to be stuck together in clusters by ____________________ rather than moving completely freely. The clusters of molecules tend to be larger and to last longer when the temperature is ____________. Smaller groups of molecules tend to form when the temperature is ____________ and the associations between water molecules are also more transient.

51. At higher temperatures, molecules are moving ____________. In order to speed up molecules, energy must be ____________. In the case of water, before a molecule can speed up, it may be necessary to break a lot of hydrogen bonds. What effect would you predict this would have on the amount of energy that is required?

52. Explain why water has a high heat capacity.

53. Because water is the solvent in cells and organisms, they too have a heat capacity that is __________. Explain why this is an advantage to living organisms.

54. A substantial amount of energy is absorbed by water when it evaporates. The amount of energy required to evaporate water is much higher than would be expected for a molecule of this size. Explain why this is so.

    The high heat of vaporization is used to great advantage by some living organisms. For example, humans produce more perspiration when the body temperature rises too high. Explain why this response has survival value.

Aqueous Solutions

Cytoplasm is the gelatinous material that fills living cells. Cytoplasm is a complex mixture of substances. Some substances are present in water as colloidal dispersions. Many other substances are dissolved in water and form a true solution. We will focus on the properties of solutions in the next few sections. Water is the solvent for cytoplasm. A solvent is the substance in which other materials dissolve. A solute is a material that dissolves in the solvent. When the solute becomes dissolved in the solvent, a solution is formed. In a solution, the solute molecules and ions are distributed evenly through the solvent.

Questions

55. Sugar water is an example of a ____________. In this case, the solvent is ____________, and the ____________ is sugar.

56. Complex solutions contain many solutes. Seawater is an example of a complex solution. In this case, the solvent is ____________, and the solutes include many different ____________.

57. In sugar water, the molecules of sugar are ______________________________ and move about independently through the solution. When the temperature is high, the molecules on average move ____________ than when the solution is at a cooler temperature.

58. Cytoplasm is a colloidal dispersion (due to the presence of proteins) and is also a ____________ aqueous solution containing many dissolved ____________. The solutes include many organic molecules, dissolved gases, and various dissolved salts. Molecules and ions of the solutes ____________.

59. Water molecules are polar ________ nonpolar _______ molecules. Materials that dissolve in water also tend to be polar ________ nonpolar ________ because the partial charges on the solute molecules are then attracted to the partial charges (polar regions) of opposite sign in the water molecules.

60. Materials that are ____________ do not tend to dissolve in water, because they have no charges or partial charges that can form attractions with the polar regions of water. Give an example of a substance that is not water soluble, and explain why you made this choice.

61. Large molecules, such as some proteins, become stably and evenly dispersed in water. The result is a ____________. How does the presence of dispersed proteins affect the consistency of the resulting mixture?

Most compounds that have polar molecules are soluble in water. Most compounds that contain ionic bonds are also water soluble. Substances that dissolve in water are described as "hydrophilic" (water loving). Nonpolar molecules such as hydrocarbons and lipids are usually quite insoluble in water. Substances that do not dissolve in water are described as "hydrophobic" (water fearing). These guidelines predict whether or not a substance will be soluble in water in the majority of cases, although there are exceptions.

Questions

62. Common table salt, with the formula NaCl, contains ___________ bonds. It is ___________ in water.

63. Glucose has the formula $C_6H_{12}O_6$. It is a small organic molecule with a carbon backbone. Sugar ______ very soluble in water, in part because the sugar molecule is polar. Also, dissolved sugar molecules can form ____________ bonds with water. Does this tend to increase or decrease the solubility of sugar? Explain.

64. Fill in the table below. First, determine whether the compound contains ionic or covalent bonds. If it contains covalent bonds, predict whether the molecule will be polar or nonpolar. Then predict whether the compound will be soluble in water. The first example is worked for you.

Compound	Type of Bond and Molecule	Water Solubility
$MgCl_2$	Ionic bonds	Water soluble—hydrophilic
Sucrose—$C_{12}H_{22}O_{11}$		
Ammonia—NH_3		
Propane—C_3H_8		
NaF		
CH_4		

65. Fats are not very soluble in water. The fat molecules are likely to be mainly ___________ in character and can be described as ____________.

66. Would you predict that nonpolar molecules such as methane and fats would be more likely to dissolve in water or in oil? Explain your answer.

Compounds such as NaCl contain many ions. Ions with opposite charges are attracted to each other, leading to the formation of ionic bonds within crystals of the compound. When NaCl dissolves in water, the individual ions dissociate and move about independently in the solution. Pure distilled water does not conduct electricity well. After ions dissolve in water, the solution becomes a good conductor of electric current. This is due to the presence of the mobile charged particles, the ions, that have been released in the solution. Solutes that release ions in solution are therefore called "electrolytes."

Questions

67. When NaCl dissolves in water, it dissociates, releasing ____________ ions and ____________ ions. The ions are charged and can move freely in the solution. Because of the presence of mobile charged particles, the solution ____________ conduct an electric current.

68. A solute that produces a conducting solution after it dissolves is called an __________________.

69. A dissolved ion becomes hydrated when a cluster of water molecules surrounds it. The sodium ion has a ____________ charge. Which part of the water molecules will be closest to the sodium ion? ______________________. Explain your answer.

70. Chloride ions have a ____________ charge. When they are hydrated, the ____________ part of the water molecule is closest to the ion.

71. In the space below, sketch a hydrated Na^+ and a hydrated Cl^-.

72. Predict whether each of the following molecules would be an electrolyte and explain why.

Glucose ____

$CaCl_2$ ____

Methane ____

KF ____

O_2 ____

Molecules such as sugar contain polar covalent bonds and are polar molecules. When crystals of sugar dissolve in water, the molecules of sugar move about independently of other sugar molecules. The covalent bonds that link the atoms together in one sugar molecule do not dissociate. No ions are released. The sugar molecules are not charged. A sugar solution does not conduct a current. Solutes of this type are nonelectrolytes.

Questions

73. Ethyl alcohol, C_2H_6O, is a molecule held together by ____________ bonds. It is a polar molecule and ______ soluble in water. It does not release any ions and is a __________________.

74. Glycerol is an organic compound with the formula $C_3H_8O_3$. The bonds that link atoms together in the glycerol molecule are ____________. The molecule is hydrophilic and ____________ dissolve in water. Ions ____________ released, and this solute is a __________________.

75. KCl is a compound that contains potassium linked to chlorine by an ____________ bond. When it dissolves in water, ions ____________ released, and the solution is an __________________.

76. Fill in the table below. For each compound, determine whether the molecule is soluble in water. For those that are, determine whether they are electrolytes or nonelectrolytes. Explain your reasons for each answer.

Compound	Water Solubility	Electrolyte or Nonelectrolyte
C_5H_{12}		
CH_4O		
$C_5H_{10}O_5$		
KBr		
C_4H_{10}		
$Ca(OH)_2$		

Acids, Bases, and Salts

The electrolytes are divided into three groups: acids, bases, and salts. Acids are compounds that can release hydrogen ions in an aqueous solution. Such compounds usually contain hydrogen linked to an element or group drawn from the right-hand portion of the periodic table. (Hydrogen is thus bonded to very electronegative elements, and its linkage to them will be ionic or strongly polar.) Four common acids are: HCl—hydrochloric acid, H_2SO_4—sulfuric acid, HNO_3—nitric acid, and H_3PO_4—phosphoric acid.

Questions

77. Acids are ____________ because they release ions in solution. The ions are mobile and are ____________. A solution containing mobile charged particles ______ conduct a current. Acids release ______ ions when they dissolve in water and dissociate.

78. The formulas for the four acids listed above all begin with the element ____________. Hydrochloric acid has ______ hydrogen(s) and can release ______ hydrogen ion(s) into solution. Sulfuric acid has ______ hydrogen(s) and can release ______ hydrogen ion(s) into solution. Nitric acid has one hydrogen and can release ______ hydrogen ion(s) into solution. Phosphoric acid has ______ hydrogen(s) and can therefore release ______ hydrogen ion(s) into solution.

79. Before dissolving and dissociating into ions, are the acidic compounds charged?

    Yes ______ No ______

    When HCl dissolves, it releases a positively charged ______ ion and a ____________ ion.

80. Sulfuric acid releases ______ hydrogen ion(s) in solution, each with a charge of ______. A total of ______ positive charges are released. The remainder of the molecule stays together as a unit and is called the sulfate group. The charge on the sulfate group is ____________.

81. Complete the table below. For each acid, fill in the number and kind of positive ions released and the number and kind of negative ions released. Remember that the molecule is neutral prior to dissociation in water.

Acid	Positive Ions Released	Negative Ions Released
HCl		
H_2SO_4		
HNO_3		
H_3PO_4		

One group of bases (also called alkalis) includes electrolytes that can release the OH^- ion (the hydroxide ion) in an aqueous solution. Four such bases are: NaOH—sodium hydroxide, KOH—potassium hydroxide, $Ca(OH)_2$—calcium hydroxide, and NH_4OH—ammonium hydroxide.

Questions

82. NaOH is a strong inorganic base called sodium hydroxide (it is the major component in lye). When it dissolves in water, it releases ____________ ions and ____________ ions. The bond between Na^+ and OH^- is ____________. NaOH is classified as a base because it releases ____________ ions in solution.

83. $Ca(OH)_2$ is another base called ____________. What is the meaning of the subscript 2 at the end of the molecule?

84. When $Ca(OH)_2$ dissolves in water, it releases a calcium ion with a charge of ____________ and ____________ hydroxide ions. Each hydroxide ion has a charge of ____________. The original calcium hydroxide is neutral. It contains charged ions called ____________ linked by ____________ to the calcium ion. Because the ions are mobile charged particles that separate in an aqueous solution, $Ca(OH)_2$ is an ____________.

85. The bases listed above contain the hydroxide ion that is listed ____________ in the chemical formula for the molecule. Preceding this, there is an ion with a ____________ charge.

86. Fill in the table below. For each base, fill in the number and kind of positive ions released and the number and kind of negative ions released.

Base	Positive Ions Released	Negative Ions Released
NaOH		
KOH		
$Ca(OH)_2$		
NH_4OH		

Salts are electrolytes that release ions other than H^+ or OH^- in solution. Common table salt, NaCl, is one example. More complex examples would include such compounds as $CaSO_4$ and NH_4Cl. Many salts are present in dissolved form in seawater, cytoplasm, and the body fluids of many organisms. Several ions derived from salts play important roles in the biochemistry of living cells, and the internal concentrations of these ions are often carefully regulated.

Questions

87. NaCl is an electrolyte that is classified as a ____________ because it releases ions other than H^+ or OH^-. When NaCl dissolves in water, it releases ______ ions and ______ ions.

88. $CaSO_4$ is another ____________ that releases ____________ ions with a charge of ______ and ____________ ions with a charge of ______ when it dissolves in water.

89. Salts normally contain a ____________ ion listed first in the chemical formula followed by a ____________ ion that is listed second.

90. The ammonium ion has a charge of ______. The phosphate ion has a charge of ____________. Can these ions form an ionic bond? Explain your answer.

    Yes ______ No ______

91.* A neutral compound must have equal numbers of positive and ____________ charges within it. Because a phosphate group has a charge of ______ , it is necessary to combine ______ ammonium ions with it to create a neutral compound. The name of this compound is __________________. Write the chemical formula for this compound. __________________

92.* Potassium ions have a charge of ______. Phosphate ions have a charge of ______. To form a neutral compound from these components, it would be necessary to combine ______ potassiums with each phosphate. The name of this compound would be __________________. Write the chemical formula for this salt. __________________

93.** Could a salt form from calcium ions and phosphate ions? Explain your answer.

    Yes ______ No ______

    If you answered yes, work out the formula for the salt, and explain all the steps of the process.

The pH Scale and Neutralization

The pH scale is a numerical scale that ranges from 0 to 14. The pH is a measure of the acid or base concentration of a solution. A pH of 7 is neutral. A pH below 7 indicates an acidic solution, and a pH above 7 indicates a basic solution. (A base is also called an alkali, and another term for a basic solution is an alkaline solution.) The pH scale is logarithmic (see the formal definition below).

The pH Scale

0	1	2	3	4	5	6	7	8	9	10	11	12	13	14

More Acidic ← Neutral → More Basic

pH is defined as: $-\log [H^+]$ (this is the negative logarithm to the base 10 of the hydrogen ion concentration).

Questions

94. Freshly distilled water that has no solutes dissolved in it would have a pH of ____________, indicating that the liquid is neither basic nor acidic.

95. The pH in most living tissue is near neutral, but may not be exactly pH 7. For example, the pH in the blood of a healthy human is 7.4. The blood is thus slightly ____________.

96. The pH of three solutions is measured. The values obtained are pH 6.1, 3.2, and 8.8. Which of these solutions is/are acidic ____________? Which has the highest concentration of H^+ dissolved in it? Explain.

97. Juice squeezed from a lemon tastes sour or tart and feels clean on the hands. Lemon juice is ____________ and has a pH ____________. The characteristics of lemon juice are properties of many acids.

98. Some bases feel slippery or slimy and taste bitter. They also have pH values that are ____________.

99. The gastric glands in the stomach produce a secretion containing hydrochloric acid. The pH in the stomach is therefore ____________.

100.*If you dissolve hydrochloric acid in distilled water, the pH will be ____________. If you prepare a second solution that is tenfold more concentrated, how will the pH of the second solution compare to the first? Explain.

An acid and a base that are combined will react. The chemical reaction produces water and a salt. This type of reaction is called a neutralization reaction. An example is shown below.

$$NaOH + HCl \rightarrow H_2O + NaCl$$

Questions

101. The following questions pertain to the reaction shown above: Which reactant is a base? ____________ Which reactant is an acid ? ____________ Which material is the salt? ____________

102. The pH of an NaOH solution would be ____________, and the pH of an HCl solution would be ____________.

103. If NaOH and HCl are mixed and reacted in a 1:1 ratio, the pH after the reaction is over would be ____________. Explain your answer to this question.

104. The products of the reaction of an acid with a base are ________________.

105. $Ca(OH)_2$ is a ____________ because it releases ____________ in an aqueous solution. H_2SO_4 is an ____________ because it releases ____________ ions in an aqueous solution. If $Ca(OH)_2$ and H_2SO_4 are combined, a ____________ reaction will occur. The products will be ____________ and a ____________.

106. One of the products of a neutralization reaction is always ____________. It always forms because OH^- is always available from the ____________, and ____________ is always available from the acid. When ______ and ______ combine, water is formed.

107. The ions that do not end up as part of the water molecule combine to form a ____________. Which reactant provides the positive ion that ends up in the salt? ____________ The negative ion that ends up in the salt is provided by the ____________.

108. What positive ion is provided by $Ca(OH)_2$ in an aqueous solution? ______ H_2SO_4 releases the negatively charged polyatomic ion ____________ in an aqueous solution. If calcium hydroxide and sulfuric acid are combined, a ____________ reaction occurs, producing the products ____________ and ____________. The particular salt that is formed has the formula ____________ and is called ________________.

Solving neutralization problems is easier if a systematic approach is used. For each acid and base, work out the ions that will be released in solution. The hydroxide ions from the base and the hydrogen ions from the acid will always react to form water. The remaining ions will combine to form the salt.

Questions

109. Under what circumstances will a neutralization reaction occur?

110. KOH is a ______ that releases ______ ions and ______ ions in an aqueous solution.

111. H_2SO_4 is an ______ that releases ______ ions and ______ ions in an aqueous solution.

112. When KOH and H_2SO_4 are combined, would a chemical reaction occur, and if so, what type?

113. The ______ derived from KOH and the ______ derived from H_2SO_4 combine to form water. The remaining ions, ______ from the base and ______ from the acid, react to form a ____________. The formula for this compound is ____________.

114. Complete the equation below by writing in the correct products.

$$KOH + H_2SO_4 \rightarrow \_\_\_\_\_\_\_\_\_\_ + \_\_\_\_\_\_\_\_\_\_$$

115. Is the equation you completed in question 114 balanced? Explain.

     Yes ______ No ______

116. How many potassiums are present in the products of the reaction (question 114)? ____________ There is/are ____________ potassium(s) present in the reactants for this reaction. A coefficient of ____________ before ____________ will achieve balance for this element. Rewrite the equation with the new coefficient. __________________

117. Explain why the equation as rewritten in question 116 is still not completely balanced.

     How many hydrogens are present in the reactants (question 116)? ______ There are ______ hydrogens in the products for this reaction. These numbers can be equalized by placing a coefficient of ______ before the formula for water.

118. Rewrite the equation for this neutralization reaction, and state whether it is balanced.

*Intermediate Difficulty

Neutralization problems should be solved systematically. First, identify the acid and determine what ions it will release in solution. Do the same for the base. Next, determine what products will form. All the H^+ derived from the acid and all the OH^- derived from the base will combine to form water. (Do not worry if the number of H^+ and the number of OH^- are different. This is corrected later during balancing.) Combine the remaining ions to form a neutral salt. Remember that the number of positive charges and the number of negative charges in the compound must be the same. Once the products have been determined, the equation can be balanced using the methods you learned previously.

Questions

119. Nitric acid has the formula HNO_3. When it dissolves in water, it releases ______ and ______.

120. Ammonium hydroxide is the base NH_4OH. It releases ______ and ______ ions in solution.

121. When nitric acid and ammonium hydroxide are combined, a ________________ reaction can occur. The products of this reaction would be ____________ and ____________.

122. Write and balance the equation for the chemical reaction in question 121.

123. When sulfuric acid and ammonium hydroxide are combined, the products are ____________ and ____________.

124. Write a balanced equation for the reaction in question 123, and explain all the steps in the balancing process.

125. If equal amounts of acid and base are combined, a ________________ reaction occurs, and the pH at the end of the process is ____________.

Buffers

A buffer is a mixture of substances that stabilizes the pH of a solution. A buffer is a mixture of a weak acid and a weak base. A weak acid is an acid that is partially dissociated. A weak base may be a substance that can accept hydrogen ions. (Weak bases usually do not release hydroxide ions.) When strong acid or base is added to a buffered solution, the pH changes very little, much less than in an unbuffered solution. The pH is quite constant in most parts of living organisms, largely because several buffers are present in the fluids that fill and surround cells.

Questions

126. Some sulfuric acid is added to pure distilled water. The pH will ____________ substantially. A set of buffering compounds is dissolved in pure distilled water. The same amount of sulfuric acid is added as in the first case. The pH will change ____________ than in the pure distilled water (which is not buffered).

127. A buffer ____________ a change in pH due to the addition of a base or an acid.

128. Acids such as sulfuric and phosphoric are strong acids that release ______ ions in solution. The hydrogen ions are linked by ______ bonds to an electronegative ion in the acid molecule. Ions move into solution ____________, and the solution is therefore an ________________.

129. Carbonic acid has the formula H_2CO_3. It is a weak acid. Carbonic acid partially dissociates to release hydrogen ions as shown in the equation below.

$$H_2CO_3 \rightleftarrows H^+ + HCO_3^-$$

A double arrow connects the reactants and products for this reaction. What is the meaning of this symbol?

130. When ions are linked by ionic bonds, they ____________ in an aqueous solution. When elements are linked by covalent bonds, they ____________ in an aqueous solution. Carbonic acid partially dissociates. Explain why this occurs.

131. The bases considered previously are strong bases, because they release the ______ ion in solution and because they dissociate completely. Some compounds function as bases but do not meet this definition. A base can be defined more generally as any substance that can accept hydrogen ions. Consider the carbonic acid dissociation reaction.

$$H_2CO_3 \rightleftarrows H^+ + HCO_3^-$$

Carbonic acid is a weak acid, because it partially dissociates and releases ______ in solution. The HCO_3^- group is called bicarbonate. Bicarbonate is a weak base, because it can ____________ hydrogen ions and react to form carbonic acid; the reaction is ________________.

*Intermediate Difficulty

A buffer is a mixture of a weak acid and a weak base. It stabilizes pH because it can react with a strong base or acid that is added to the buffered system. The weak base can react with strong acid, and the weak acid can react with a strong base, thus neutralizing them.

Questions

132. Carbonic acid is a ____________, and bicarbonate is a ____________. If carbonic acid and bicarbonate are both present, the mixture will function as a ____________ and will tend to resist a change in pH.

133. The following questions all pertain to a solution that contains a carbonic acid/bicarbonate buffer system. HCl is an ____________, because it releases ____________ ions in solution. If HCl is added to the buffered solution, the hydrogen ions from the ____________ can be accepted by the ____________, a weak base. After this reaction has occurred, there will be more ____________ in the buffered system and less ____________. The change in pH will be ____________ than in an unbuffered system.

134. NaOH is a ____________, because it dissociates completely and releases ____________ in solution. If NaOH is added to the same buffered system, the ____________ ions will be neutralized by reacting with ____________. After this reaction, there will be more ____________ and less ____________ in the buffered system. The change in pH will be ____________ than in an unbuffered system.

135. Carbonic acid is formed by the reaction represented in the following equation.

$$CO_2 + H_2O \rightleftarrows H_2CO_3$$

The double arrow between reactants and products means that ________________. At any time, there will be a mixture of carbon dioxide, water, and ____________. This is another example of a reaction that ____________ go to completion.

136. Carbon dioxide is produced by cells as they metabolize. As CO_2 is produced, some of it reacts with water and forms ____________. After the carbonic acid has formed, some of it dissociates and releases _______and ____________.

137. The carbonic acid/bicarbonate buffered system is important in living systems. In light of your answer for question 136, explain why this is so.

138. Cells produce carbon dioxide as they metabolize. They also get rid of carbon dioxide, generally at the same rate as they produce it. The concentrations of carbonic acid and ____________ should therefore be relatively constant in and around the cell.

Molecular Weight

The molecular weight (or mass) of any molecule is the sum of the weights of all the atoms in that molecule. Atomic weights or masses can be obtained from the periodic table of the elements. A sample calculation of molecular weight is shown below.

Molecular weight of water, H_2O, equals:

weight of two H atoms + weight of one O atom = 1 + 1 + 16 = 18 amu or daltons

Questions

139. The following question refers to the example above. The mass of one hydrogen atom is ______. (Round off to the nearest tenth of a unit.) The units for atomic weights or masses are ____________ or ____________. The water molecule contains ______ hydrogens, weighing a total of ____________, plus one atom of ____________, weighing ____________. The water molecule contains a total of ____________ atoms, weighing ____________. The molecular weight of water is ____________.

140. The formula for glucose is $C_6H_{12}O_6$. Calculate the molecular weight of glucose by completing the table below.

Atom	Number in Molecule	Atomic Weight	Total Weight
C	6	12	72
H			
O			

Molecular Weight =

141. The formula for ammonium phosphate is ____________. Design a table similar to the one above, and calculate the molecular weight of this compound.

Chemists have defined a very useful unit called the mole. Moles and molarity, a concentration term based on the mole, have many important applications in aqueous chemistry. A mole is defined as a gram molecular weight, that is, the molecular weight expressed as grams. For example, a mole of water is 18 grams of water, and a mole of glucose is 180 grams of glucose. A mole of any compound always contains the same number of molecules. The number of molecules in a mole is called Avogadro's number, which has a value of 6.022×10^{23}.

Questions

142. The molecular weight of water is ______. One mole of water has the same numerical value of ______, but instead of daltons, the unit is now ______. One molecule of glucose weighs ______ daltons, and one mole of glucose is ______ of glucose.

143. Define the term molecular weight. ____________
Define the term mole.

Explain the relation between molecular weight and a mole. How are they similar and how do they differ?

144. One molecule of water weighs ______ daltons, and one molecule of glucose weighs ______ daltons. The glucose molecule weighs ____________ (what multiple?) more than the water molecule. When you have 18 daltons of water and 180 daltons of glucose, you have ______ molecule(s) of each, and the number of molecules is ____________.

145. Ten molecules of water weigh ____________, and ten molecules of glucose weigh ____________. The number of molecules is ____________, and the weight of the glucose molecules is ____________ that of the water molecules.

146. One hundred molecules of water weigh ____________, and 100 molecules of glucose weigh ____________. The number of molecules is ____________, and the weight of the glucose molecules is ____________ that of the water molecules.

147. You are given a certain weight of water, weight X. To have the same number of glucose molecules, you must obtain a weight of ____________, because glucose molecules weigh ____________ more than water molecues.

148. A mole of water weighs ______, and a mole of glucose weighs ______. The number of molecules in the mole of water will be ____________ the number of molecules in the mole of glucose. Explain.

149. A mole of water contains ____________ molecules. A mole of glucose contains ____________ molecules.

Molarity

Molarity is a measure of concentration. The molarity is the number of moles of solute dissolved per liter of solution. A 1.0M (one molar) solution has one mole of solute dissolved in a liter of solution. Equimolar solutions have the same concentration of solute molecules. This property is very useful when it is necessary to combine defined proportions of molecules for a chemical reaction.

Questions

150. A liter of a 1.36M solution of glucose has ______ moles of glucose dissolved in the solution.

151. A 0.72M solution of ammonium phosphate has ______ moles of ammonium phosphate dissolved per liter of solution.

152. The unit of molarity is ______ per ______. The symbol M represents this unit.

153. A glucose solution that is 0.46M is equimolar to a ______ solution of sucrose. The equimolar glucose and sucrose solutions contain ____________ concentration of solute molecules. Ten milliliters of the glucose solution contains the same number of solute molecules as ______ of the sucrose solution.

154. One mole of glucose weighs ______ grams. Carefully weigh out one mole of glucose, ______ grams, and dissolve it in distilled water. After the solute has fully dissolved, add water until you have exactly one liter of solution. The molarity of this glucose solution is ______. Prepare a label for your solution that says ______, and fasten it to the storage bottle.

155. Sucrose is a sugar that has the formula $C_{12}H_{22}O_{11}$. Calculate the molecular weight of sucrose, and then explain how to prepare a 1M solution of this sugar. Show your work.

156. One mole of glucose weighs ______ grams and contains ________ molecules of glucose. One mole of sucrose weighs ______ grams and contains ________ molecules of sucrose. The number of molecules of glucose in one mole is ______ to the number of molecules of sucrose in one mole.

157. A liter of 1M glucose is prepared by using ______ grams of glucose, which is ______ mole. A liter of 1M sucrose is prepared by using ______ grams of sucrose, which is ______ mole. An ______ number of solute molecules are dissolved in ______ volumes of solution. The concentration of glucose and the concentration of sucrose are ______. The number of molecules of glucose in 3 ml of 1M solution is equal to the number of molecules of sucrose in ______ of 1M solution, because the molarities are equal.

It is often convenient to prepare solutions of a variety of molarities and volumes. The following formulas permit you to calculate the amount of solute needed to prepare solutions if the volume and molarity are specified.

Formula 1	Molarity × Volume = Number of Moles
Formula 2	Number of Moles × Mass of 1 Mole = Number of Grams Required

Questions

158. The units of molarity are ______ per ______. When this unit is multiplied by a volume, the answer has units of ______.

159.*You wish to prepare 500 ml of a 0.50M glucose solution. How many moles of glucose will be needed? Show your work.

160. The units for the mass (or weight) of one mole are ______ per ______. When this unit is multiplied by number of moles, the answer has units of ______.

161.*How many grams of glucose are required to prepare the solution requested in question 159? Show your work and include units.

162.*You are now ready to prepare the solution requested in question 159. First weigh out ______ grams of glucose and dissolve it in pure ____________. When the glucose is completely dissolved, add distilled water to reach a final volume of ______. Pour the completed solution into a storage bottle, and attach a label that says ______.

163. You are now preparing for an experiment, and you remove a 5-ml sample of the glucose solution from question 162. The concentration of this glucose is ______.

164.**Calculate how many grams of glucose would be required to prepare 400 ml (remember that 1 liter = 1000 milliliters, or ml) of a 0.36M solution. Include units on all calculations.

165.* Explain how you would prepare the solution requested in the previous question.

**Advanced Difficulty

Questions

166. The structural formula for the amino acid called serine is shown below. The chemical formula is ____________. In the space below, set up a table and calculate the molecular weight of serine.

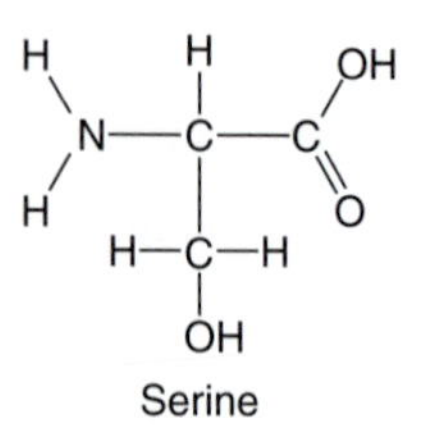

Serine

167. How many grams of serine would be required to prepare 650 ml of a 0.8M solution? Show your work.

168. Explain how you would prepare the solution requested in question 167.

For each of the following problems, calculate how much solute would be needed, and explain how the solution would be prepared.

169. 100 ml of a 1.5M solution of glucose.

170. 750 ml of a 0.2M solution of sucrose.

171. 95 ml of a 0.04M solution of serine.

Dilutions

Concentrated stock solutions are often prepared for chemicals that are used frequently. Solutions of a variety of concentrations can be prepared by diluting the stock solution. The formula below can be used to calculate how to carry out such a dilution.

$$\text{Initial concentration} \times \text{Initial volume} = \text{Final concentration} \times \text{Final volume}$$

or:

$$IC \times IV = FC \times FV$$

Questions

172. You have a stock solution of 4M NaCl. You wish to prepare 200 ml of a 2M NaCl solution. The formula above can be used to solve this problem. The initial concentration refers to the concentration of the stock solution, that is ____________. The final concentration refers to the concentration of the diluted sample and is ____________. The final volume is the volume of diluted solution requested and is ____________.

173. How much of the stock solution should be used to prepare the dilution requested in question 172? (Hint: this is the initial volume.) Show your work.

     How much water must be added to prepare the requested dilution?

174. Explain how you would prepare the following solution. Show all calculations. How would you prepare 500 ml of a 0.25M solution of glucose from a 3M stock solution?

175. You have 20 ml left of a 1M alanine solution. What is the maximum amount of 0.015M solution that can be prepared from this? How would you do this?

176. Explain why the formula $IC \times IV = FC \times FV$ is valid. In other words, why are the two sides of this equation equal?

Aqueous Chemistry: An Overview

You have now learned about the properties of water, an exceptional molecule that is essential to life as we know it. Artists represent the shape of the water molecule in various ways. One of the most common diagrams is on page 160. This version of the water molecule shows it as an approximately triangular molecule. The two unbonded pairs of electrons are not represented in this diagram. If they were included, it would be more obvious that the water molecule is not a triangle, but rather a three-dimensional tetrahedron. The oxygen end of the molecule has a partial negative charge, and the hydrogen end of the molecule has a partial positive charge. This polarity is due in part to the unbonded electron pairs, and also to the fact that oxygen is electronegative and the electron cloud is displaced toward it.

Water molecules can form hydrogen bonds to each other and to many other substances as well. A simple way of imagining hydrogen bonds is to think of them as conferring on water a kind of chemical stickiness. Water in the liquid phase has clusters of water molecules associated by hydrogen bonds. As water warms, the clusters are smaller, on average, and more transient. Many of the unique properties of water are a consequence of its ability to form hydrogen bonds.

Water is an excellent solvent, although it does not dissolve all compounds. Substances that are soluble in water usually contain particles that have a full or partial charge. The charges in the solute can then interact with the partial charges on water. Compounds that have ions linked by ionic bonds are frequently water soluble. Water molecules surround the ions and escort each one into solution separately. The diagram on page 160 shows ions entering an aqueous solution in this manner. Once in solution, the ion is hydrated, or surrounded by a shell of oriented water molecules. The attractions between the polar regions of the water molecules and the charges on the ions stabilize these structures. Polar molecules such as sugars are also water soluble. These molecules enter solution as a unit and form hydrogen bonds with water. Nonpolar substances such as hydrocarbons tend to be insoluble in water and are referred to as hydrophobic. They do not tend to dissolve in water, because no hydrogen bonds or hydration shells can be formed with the water molecules.

Cytoplasm is the material that fills the living cell. Cytoplasm is both a complex aqueous solution and a colloidal dispersion. Many materials are dissolved in cytoplasm and form true solutions. Some large molecules, such as protein, are present in cytoplasm but are not truly dissolved. Rather, these components are stably dispersed in the cytoplasm. The substances in colloidal dispersion are mostly responsible for the gelled consistency of the cytoplasm.

Let us now move to the submicroscopic world of the water molecules and take a close look at a complex solution such as seawater. As we enter this world, we see hundreds of triangular electronic clouds. Each water molecule has two additional electron pairs sticking out in front and projecting to the back, almost as if the water molecule were wearing a sideways hat of electrons, perched on the oxygen. As we try to push through the water, we find that the medium is viscous. The water molecules are stuck together and pushing through or around them is therefore difficult. As we make laborious progress through the watery medium, we see many other molecules and ions all surrounded by water.

The fossil record shows that life began in a medium such as this. In the primordial eras of our earth, chemical evolution formed compounds that dissolved in the early seas. In selected locations, this could have led to the formation of a complex organic soup, a rich medium for the development of early life-forms. Now, many organisms, including humans, have left the water behind and live on land instead. Yet, in a very real sense, we have not left the seas but carry our own small one with us. The internal fluids of our body resemble the composition of the early seas, and elaborate physiological mechanisms are dedicated to maintaining the composition of our internal fluids. The next time you enjoy a cool glass of water, think of the tiny tetrahedral molecules within it. They provide the medium in which the biochemistry of your life is conducted and are the very stuff of the cells.

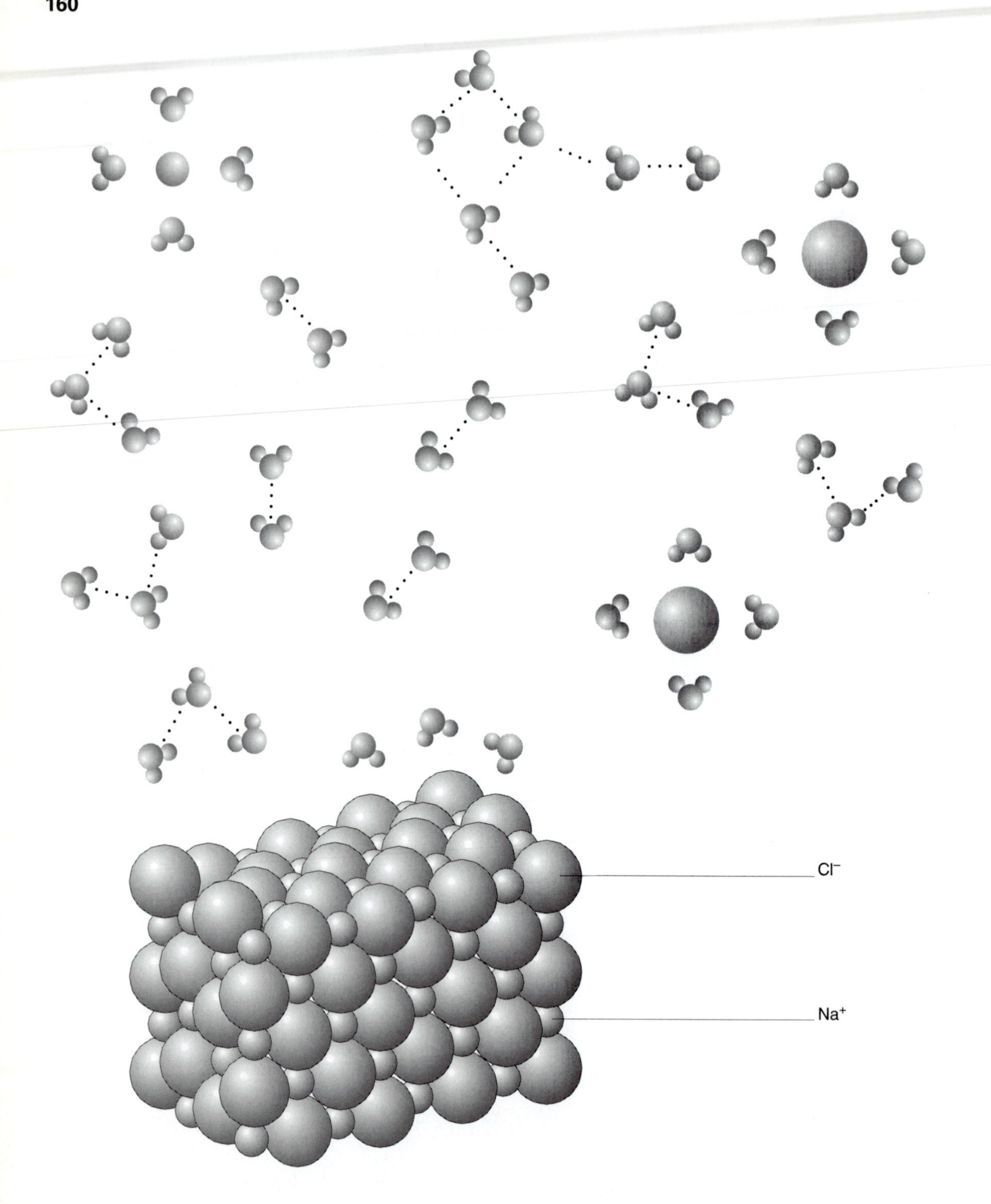

A Sodium Chloride Crystal

Answers

1. water

2. water (dissolves the other materials); solutes (materials that dissolve); solvent; solutions.

   Note: Water is correctly described as an excellent solvent, because many substances dissolve in it. Water is sometimes referred to as the universal solvent. This is an incorrect term, however, because some substances are not soluble in water. Oily and fatty compounds are among those substances that are not water soluble. Molecular shape, the presence or absence of ionic charges, and a property called molecular polarity all affect the water solubility of a substance. Molecular shape and polarity will be presented in future sections.

3. table salt (NaCl) and sugar (solutes); water (solvent). Both a saltwater solution and a sugar solution would be clear and transparent.

   Note: True solutions are transparent and also stable over time. That is, the solutes do not settle out of solution as time passes.

4. Seawater is one complex aqueous solution. The solvent is water, and several salts are the solutes that are dissolved in the water. Many organisms live in the sea, and the fossil record and other evidence strongly supports the idea that life originated in the earth's primitive seas. Cytoplasm is also a complex aqueous solution, because water is the solvent and there are many dissolved solutes. (This, however, is not a full description of cytoplasm, because its structure is more complex.) Cytoplasm fills the living cell.

5. proteins; colloidal dispersion; translucent.

6. Cytoplasm is an example of a colloidal dispersion. There are many large molecules, especially proteins, that are stably dispersed in water. The gelled consistency of cytoplasm is largely due to these colloidal components.

7. water.

   Note: Since cytoplasm is both a solution and a suspension that uses water as its solvent, this is a reasonable guess. Water is the most abundant compound in living organisms, making up, for example, about 70% of the human body. Water is the solvent in all life-forms known on earth.

8. 4.

Note: There are four faces in a tetrahedron. One face forms the base, and the other three faces form the sides, meeting at the top. The overall shape is a pyramid.

9.*Each face or surface is a triangle. Each of the four faces is exactly the same size and shape in a regular tetrahedron.

10. Carbon is at the center. Carbon is located inside the tetrahedron and is at the center of the tetrahedral shape. Hydrogens are bonded to carbon, the central atom. They are located at the four points or apexes of the tetrahedral shape.

11. A covalent bond connects each hydrogen to the central carbon. In the flat projection formula, each covalent bond is represented by a dash.

12. A shared pair of electrons (covalent bond); electrons (negative charge); repel; the same.

13.**Each covalent bond in the methane molecule contains a pair of electrons, which are negatively charged. Like charges repel each other, and the electrons push away from each other as far as possible. The tetrahedral shape forms, because this shape allows the electrons to maximize their distance from each other. This is the shape that permits the four pairs of electrons to be farthest apart.

14. nitrogen (center of ammonia); oxygen (center of water); inside and at the center.

15. hydrogen; hydrogen. The hydrogens are located at two (water), or three (ammonia) of the points, or apexes, of the regular tetrahedron.

16. There is one pair of unbonded electrons in the valence shell of nitrogen (in the ammonia molecule). Two (pairs of unbonded valence electrons in water); In ammonia, the unbonded pair of electrons is located at an apex or point of the regular tetrahedron. In water, the two pairs of unbonded electrons are located at two of the apexes of the regular tetrahedron.

17.*Ammonia and water assume a tetrahedral shape, because all four pairs of valence electrons can maximize their distance from each other in this configuration. The electrons are negatively charged and repel each other whether or not they participate in chemical bonding.

18.*90°.

19.*tetrahedral; about 109°; 109°; 90°; mutually repulsive; as far apart from each other as possible.

20.**Water is not a linear molecule, because the central atom of oxygen has four pairs of valence electrons. All these electron pairs, whether bonding or nonbonding, are negatively charged and repel each other. The electrons can get farther apart in a tetrahedron than in any other shape. The bond angle in a tetrahedron is the greatest angle that can form between four mutually repelling pairs of electrons.

21. Covalent bonds.

22.*The carbon-to-hydrogen bonds are slightly polar. The electronegativity of carbon is 2.5, and that of hydrogen is 2.1 (see chapter 3 if you need to review the topic of electronegativity). When two atoms with different electronegativities become linked by covalent bonds, the bond is polar. When the difference in electronegativities is small, as is the case here, the bond is only slightly polar.

23.*carbon (electrons displaced toward); hydrogen (electrons displaced away from).

24.*109° (bond angle in methane); are all (the same); hydrogen (occupies all four bonding positions); methane is (symmetrical).

25.**polar; center of the molecule, on the carbon atom; center of the molecule, on the carbon atom.

26.**The molecule as a whole is symmetrical in all directions. Although the bonds involve slight separations of charge, they cancel each other out because of this symmetry.

The slight electronic displacements that result from the polar bonds are all of the same magnitude, and they offset each other because they pull in a set of opposing directions.

27. Ethane is a hydrocarbon. Hydrocarbons are generally nonpolar for the same reasons that apply to methane.

The bonds within them are only slightly polar, and the symmetrical geometry of the molecule causes these slight electronic displacements to cancel each other, because they pull in opposing directions.

28.*polar covalent (bonds linking H to O); toward oxygen, electronegativity of 3.5; away from hydrogen, electronegativity of 2.1.

29. Two pairs (nonbonding valence electrons); two of the points or apexes.

30. at two of the apexes.

31. The water molecule has the shape of a tetrahedron.

    On one side of the water molecule, there is an oxygen and two unbonded pairs of electrons. The other side of the water molecule has two hydrogens. When the water molecule is rotated, these different sides are evident. This configuration differs from methane. No matter how methane is rotated, all parts of the molecule are equivalent, because methane is symmetrical in all directions. No part can be distinguished from any other, because all four bonding positions are the same.

32. toward oxygen; away from hydrogen; oxygen (center of negative charge); oxygen; hydrogen; hydrogen (center of positive charge).

33. A polar molecule has some internal separation of charge.

    There is a slightly positive pole at one location in the molecule, and a slightly negative pole at a different location. This occurs when the electron cloud (with its negative charges) shifts toward more electronegative groups within the molecule. Since the nuclei (with the positive charges) do not move, positive charge is then centered at a different point in the molecule than negative charge. A molecule of this type is polar.

    Yes, water fits the definition of a polar molecule.

    Negative charge is centered near the oxygen atom in water, and positive charge is centered nearer the two hydrogen atoms and between them. Since these points are not the same, water has two separate poles, positive and negative, making it a polar molecule.

34. The δ^+ identifies the positive pole in the water molecule. It shows there is a partial positive charge in the vicinity of the hydrogens that is caused by shifting the negatively charged electron cloud away from hydrogen and partially exposing their positively charged nuclei. The δ^- signifies the negative pole within the water molecule. There is a partial negative charge near oxygen, because oxygen is an electronegative element and pulls on the electron cloud.

35. polar.

    Each oxygen atom pulls some of the electron cloud toward itself. Because of the shape of the carbon dioxide molecule, the oxygens pull in opposite directions. The displacements of the electron cloud offset each other. It is somewhat like a tug of war with equally matched opponents. Because the displacements cancel each other, carbon dioxide is a nonpolar molecule. Carbon dioxide is a linear molecule, rather than tetrahedral, because it contains double bonds. The presence of double bonds alters the geometry of the molecule. One double bond points in one direction, and the other is directly opposite to the first. A molecule with just two double bonds, therefore, forms a straight line.

36. polar covalent; oxygen; polar.

    Water is a polar molecule, because it contains polar covalent bonds and because of its shape and geometry. Unlike carbon dioxide, water contains no double bonds. There are four pairs of valence electrons. Two of the pairs are bonding electrons, and two pairs are nonbonding. Each of the four pairs repels the others, and the molecule has tetrahedral symmetry. The electron cloud in water is shifted toward oxygen and is not offset or cancelled by any shift in the opposite direction. There is a partial negative charge near the oxygen and a partial positive charge near the hydrogens. Since there are poles of partial charge, water is a polar molecule.

37. oxygen (partially negative pole); hydrogen (partially positive pole).

38. hydrogen; attracted.

39. Ionic.

40. dotted lines; weak.

41. The poles within the water molecules are not full charges. Within each water molecule, there is a slight displacement of the electron cloud. Electrons have not been completely transferred. The charges that have developed are partial, and are represented by the Greek letter delta, δ. The incomplete or partial charges attract each other less strongly than the complete charges on ions.

42. No.

    Within the water molecule, the hydrogen and oxygen atoms are linked to each other with polar covalent bonds. The attractions between different water molecules do not link any atoms into molecules. The hydrogen bonds are weak attractive forces between different molecules. The molecules behave as if they are sticky. These interactions are weak and easily broken. Hydrogen bonds constantly form and break as water molecules move about.

43. No.

    Sulfur has an electronegativity of 2.6, and hydrogen has an electronegativity of 2.1. Since this is a small difference, the covalent bonds that link them are only slightly polar. Hydrogen must be bonded to a more electronegative element and form a strongly polar covalent bond before it can hydrogen bond. Only fluorine, oxygen, and nitrogen are electronegative enough to meet this requirement.

44. Methane will not form hydrogen bonds.

    The molecule as a whole is not polar. Because there are no poles, one of the conditions required for hydrogen bonding is missing. In addition, the bond between carbon and hydrogen is only slightly, rather than strongly, polar. Hydrogen must be linked to fluorine, oxygen, or nitrogen (in a strongly polar covalent bond) before it can form hydrogen bonds.

45. Yes.

    Note: In an amine group (–NH_2), hydrogen is bonded to nitrogen by a polar covalent bond. The hydrogen develops a partial positive charge (δ^+). Oxygen is bonded to carbon in a carbonyl group (–C=O) by a polar covalent bond, and the oxygen develops a partial negative charge (δ^-). Because both these groups are found within proteins, hydrogen bonds can form between different protein molecules when partial charges of opposite sign are near each other.

    Hydrogen bonds can form within protein molecules if a part of the protein containing hydrogen in an amine group folds so that it is near to a partially negative carbonyl group within the same molecule. Intrachain hydrogen bonds form in this way in proteins and are very important to protein shape and function.

46. 3; more.

47. hydrogen bonds to neighboring water molecules. An associated liquid is one that contains molecules that associate in clusters by forming hydrogen bonds.

48. can; associated; cannot; is not; do not (have similar physical properties).

49. more slowly.

50. attractions between different molecules; hydrogen bonds; lower; higher.

51. faster; added. Breaking hydrogen bonds requires energy. Hydrogen bonds are weak attachments between different molecules. In effect, the molecules behave as if they are sticky. Energy is required to pull the sticky molecules apart.

52. Water has a high heat capacity. This means it takes more energy to raise the temperature of water than would be required for most other molecules of a similar size. Water has a high heat capacity, because it contains hydrogen bonds. Energy is required to break these hydrogen bonds, as well as to speed up the molecules.

53. High. The high heat capacity of water provides organisms with temperature stability. If water did not have a high heat capacity, organisms would experience more extreme and more rapid temperature changes.

54. When water evaporates, it goes into the gas phase. In the gas phase, the molecules are much farther apart than they are in the liquid phase. All remaining hydrogen bonds must be broken before vaporization can occur, requiring a substantial input of energy.

    Temperatures that are too high can damage cells and organisms. In extreme cases, death can result. The high heat of vaporization of water helps many organisms avoid this fate.

    When perspiration evaporates, heat energy is removed from the organism, helping it cool and stabilizing temperature within a safe range. It is fortunate that water has such a high heat of vaporization. If it did not, an animal would have to evaporate far more water to achieve cooling, thus risking dehydration.

55. solution; water is the solvent; the solute is sugar.

56. water is the solvent; salts.

    Note: The term "salts" includes more than the familiar table salt, NaCl. All the compounds that contain a metal ion linked to a nonmetal ion by ionic bonds are included in this category. Examples include $CaCl_2$, Na_2O, and KI.

57. distributed evenly throughout the water; faster.

58. complex; solutes; are distributed evenly throughout the water and move about independently.

59. polar; polar.

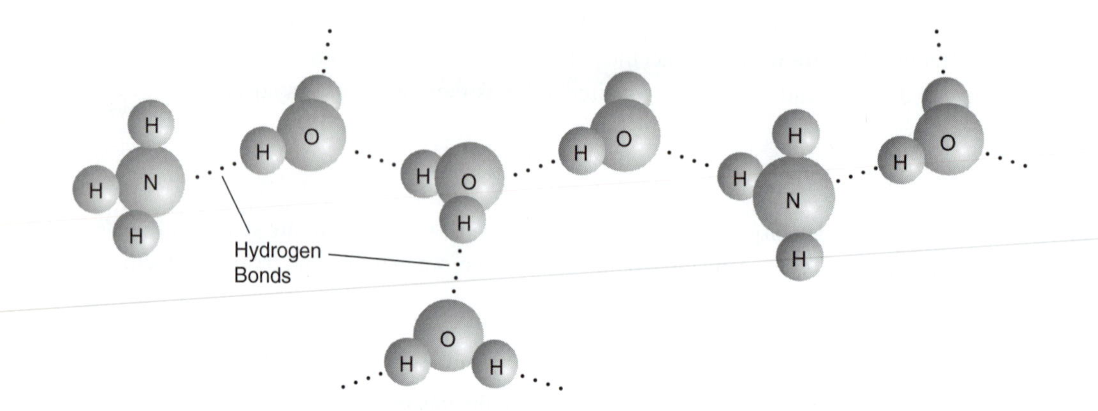

60. nonpolar. Hydrocarbons are molecules that are nonpolar. An example would be methane.

These molecules have no ions or polar regions and therefore cannot engage in hydrogen bond formation or other electronic attractions with water molecules. In living systems, fats, oils, and other lipids fall into this class. They are composed mostly of hydrocarbon, are nonpolar, and are not water soluble.

61. colloidal dispersion; Many colloidal dispersions exhibit some degree of gelling. Cytoplasm has a soft gel consistency, largely due to the proteins that are present.

62. ionic; very soluble.

63. is (soluble in water); hydrogen bonds. Hydrogen bond formation promotes the solubility of sugar in water. Substances that can hydrogen bond generally are more soluble in water than those that cannot, because the weak attractions of the hydrogen bonds help "pull" substances into solution.

64.

Compound	Type of Bond and Molecule	Water Solubility
$MgCl_2$	Ionic bonds	Water soluble—hydrophilic
Sucrose—$C_{12}H_{22}O_{11}$	Polar covalent bonds—polar molecule can hydrogen bond to water	Water soluble—hydrophilic
Ammonia—NH_3	Polar covalent bonds—polar molecule can hydrogen bond to water	Water soluble—hydrophilic
Propane—C_3H_8	Slightly polar covalent bonds—nonpolar molecule—hydrocarbon	Insoluble—hydrophobic
NaF	Ionic bonds	Water soluble—hydrophilic
CH_4	Slightly polar covalent bonds—nonpolar molecule—hydrocarbon	Insoluble—hydrophobic

65. nonpolar; hydrophobic.

66. Nonpolar molecules would be more likely to dissolve in oil, and they would not be soluble in water.

    As a rule, only polar molecules and compounds containing ions are water soluble. A simple rule of thumb is that "like dissolves like." There is therefore a good probability that nonpolar molecules will be soluble in fats, because fats are mainly hydrocarbon and are nonpolar. Hydrophobic molecules tend to stay together and avoid interactions with water.

67. sodium and chloride (ions); can (conduct a current).

68. electrolyte.

69. positive. The negative (oxygen) end of the water molecule will be nearest the sodium. Because opposite charges attract each other, the positive ion interacts with the partial negative on the oxygen end of the water molecule.

70. negative; positive.

71.

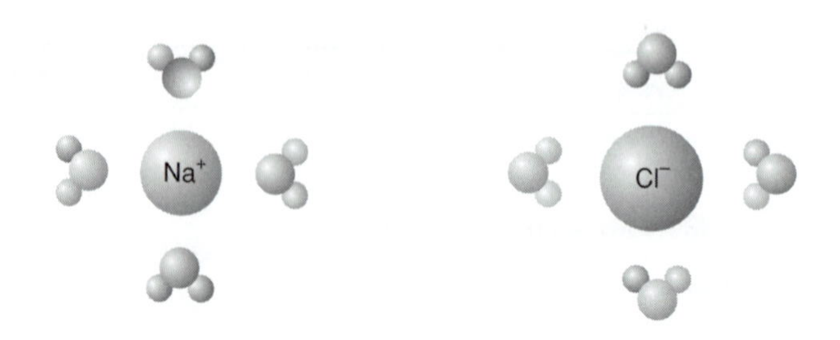

72. Glucose would not be an electrolyte. The atoms in the glucose molecule are linked by covalent bonds that do not dissociate in water. When glucose is dissolved, the glucose molecules dissolve as a unit. Glucose does not release ions. Without mobile charged particles in solution, an electric current is not conducted.

    $CaCl_2$ Calcium chloride is an electrolyte. Calcium ions are positive and chloride ions are negative. When calcium chloride dissolves in water, the ions dissociate and move about independently. The solution is then a conductor, and calcium chloride is an electrolyte.

    Methane is not an electrolyte. It has almost no solubility in water. It contains no ions and therefore could release none even if it could dissolve.

    KF is an electrolyte because it contains ions that dissociate and move independently in water.

    O_2 is not an electrolyte. The bonds linking oxygen atoms into a diatomic molecule are nonpolar and covalent. No ions are released in solution.

73. covalent; is; nonelectrolyte.

74. covalent; does; are not; nonelectrolyte.

75. ionic; are; electrolyte.

76.

Compound	Water Solubility	Electrolyte or Nonelectrolyte
C_5H_{12}	Not water soluble	Not relevant because it's insoluble
CH_4O	Soluble in water	Nonelectrolyte
$C_5H_{10}O_5$	Soluble in water	Nonelectrolyte
KBr	Soluble in water	Electrolyte
C_4H_{10}	Not water soluble	Not relevant because it's insoluble
$Ca(OH)_2$	Soluble in water	Electrolyte

C_5H_{12} and C_4H_{10} are hydrocarbons. They contain bonds that are slightly polar. Because of the symmetry of the molecules, the slight shifts of the electron clouds offset each other. The molecules as a whole are nonpolar, not very soluble in water, and hydrophobic. These are properties of other hydrocarbons as well.

CH_4O is the formula for methanol, and $C_5H_{10}O_5$ is the formula for a sugar. Both are small organic molecules containing atoms linked by covalent bonds. Bonds between oxygen and carbon are quite polar. Also, these molecules are not symmetrical in all directions; therefore, the molecules as a whole are polar, water soluble, and hydrophilic. Small organic molecules containing oxygen often show these properties. Since all the bonds are covalent, the molecules dissolve as intact units, and these solutes are nonelectrolytes.

KBr and $Ca(OH)_2$ are both compounds that contain ions. Both compounds are water soluble, releasing ions into solution. Therefore, both are electrolytes.

77. electrolytes; charged; can (conduct a current); hydrogen.

78. hydrogen; 1; 1 (hydrogen in HCl and released by it); 2; 2 (hydrogens in sulfuric acid and released by it); 1 (hydrogen released by nitric acid); 3; 3 (hydrogens in phosphoric acid and released by it).

79. No; hydrogen ion (H^+); negatively charged chloride ion (Cl^-).

80. 2 hydrogen ions; +1; 2 ; –2 is the charge on the sulfate group.

Note: One way to analyze sulfuric acid is to note that the bonds linking the two hydrogens to sulfate are ionic. Each hydrogen has donated its electron to the sulfate group. That leaves the hydrogens with one positive charge each and the sulfate with –2 (due to the two extra electrons that have been donated to it).

81.

Acid	Positive Ions Released	Negative Ions Released
HCl	H^+	Cl^-
H_2SO_4	$2H^+$	SO_4^{2-}
HNO_3	H^+	NO_3^-
H_3PO_4	$3H^+$	PO_4^{3-}

Note: Groups such as SO_4^{2-} (sulfate), NO_3^- (nitrate), and PO_4^{3-} (phosphate) are called polyatomic ions. Each polyatomic ion contains several atoms linked by covalent bonds. Each polyatomic ion is a unit that remains intact in solution. As a whole, the group is negatively charged and will bond to the number of hydrogens required to form a neutral compound. It is therefore easy to work out the charge on any polyatomic ion. It is exactly the amount required to equal the positive charges from the hydrogen ions present in that compound.

82. sodium (Na^+); hydroxide (OH^-); ionic; hydroxide.

83. calcium hydroxide. The subscript 2 means that there are two copies of the group that are enclosed in parentheses. The hydroxide ion is within parentheses, and therefore, two of these are present in the calcium hydroxide molecule.

84. +2; 2 (hydroxide ions); –1; hydroxide; ionic bonds; electrolyte.

85. last; positive.

86.

Base	Positive Ions Released	Negative Ions Released
$NaOH$	Na^+	OH^-
KOH	K^+	OH^-
$Ca(OH)_2$	Ca^{2+}	$2OH^-$
NH_4OH	NH_4^+	OH^-

Note: The groups OH^- (hydroxide) and NH_4^+ (ammonium) are polyatomic ions. The atoms within the polyatomic ion are linked by covalent bonds. The polyatomic ion remains intact in aqueous solution, carries a charge, and functions as a chemical unit.

87. salt; Na^+ ions and Cl^- ions.

88. salt; calcium; +2; sulfate ions (with a charge of) –2.

89. positive; negative.

90. +1; –3; Yes. Opposite charges attract each other. Because ammonium ions and phosphate ions have opposite charges, they can attract each other and form ionic bonds.

91.*negative; –3; 3; ammonium phosphate. The formula is $(NH_4)_3PO_4$.

Note: Parentheses surround the ammonium group. The subscript 3 indicates that there are three copies of the entire ammonium group (each with a charge of +1) within the ammonium phosphate.

92.*+1; –3; three (potassiums for each phosphate); potassium phosphate. The formula is K_3PO_4.

93.**Yes. Calcium forms an ion with a charge of +2. The phosphate ion has a charge of –3. Since these ions have opposite charges, they will attract each other and form ionic bonds.

To form a neutral compound, equal numbers of negative and positive charges must be present in the salt. Six is the lowest common multiple of two and three. Six negative charges will be contributed by two phosphate groups (–3 per group). Six positive charges will be contributed by three calcium ions (+2 per ion). The ratio of calcium ions to phosphate ions is 3:2. The formula for this compound is $Ca_3(PO_4)_2$. *The salt, calcium phosphate, is neutral.*

94. 7

95. basic.

96. pH 6.1 and pH 3.2. The solution with a pH of 3.2 has the most H^+ dissolved in it. The lower the pH, the greater the concentration of hydrogen ions.

pH is actually a logarithmic scale (see the definition on page 145). When the pH goes down one unit, the concentration of hydrogen ions is actually 10 times greater.

97. acidic; below 7.

98. above 7.

99. below 7, indicating the presence of an acid.

100. be below 7. The pH of the second solution will be lower than the pH of the first solution. As more hydrochloric acid is dissolved, more will dissociate and release H^+. The greater the concentration of hydrogen ions, the lower the pH will be. Since the pH scale is logarithmic, a solution that is tenfold more concentrated will have a pH that is one unit lower than the first one. If the first solution had a pH of 3, the second solution would have a pH of 2.

101. NaOH is a base; HCl is an acid; NaCl is the salt.

102. above 7; below 7.

103. 7. If an acid and a base are mixed in a 1:1 ratio, the neutralization reaction will use up all the acid and base. If there is no leftover acid or base, then the only substances present are the products, namely water and a salt. The solution is neutral, and the pH is then 7.

104. water plus some type of salt.

105. base; it releases OH^-; acid; it releases hydrogen ions; neutralization; water and a salt.

106. water; base; H^+ (available from the acid); OH^- and H^+.

107. salt. The base provides the positive ion for the salt; acid.

108. Ca^{2+}; SO_4^{2-}; neutralization reaction occurs; produces water and a salt; $CaSO_4$; calcium sulfate.

109. When an acid and a base are mixed.

110. base; OH^- ions and K^+ ions.

111. acid; hydrogen (H^+) ions and SO_4^{2-} ions.

112. Yes, they will react by neutralization, because KOH is a base and H_2SO_4 is an acid.

113. OH^- (derived from KOH); H^+ (derived from H_2SO_4); K^+ (from the base) and SO_4^{2-} (from the acid); form a salt. K_2SO_4.

114. $KOH + H_2SO_4 \rightarrow H_2O + K_2SO_4$

115. No. An equation is balanced when there is the same number of each kind of atom on the reactant and on the product side of the equation. Neither H nor K is balanced now.

116. 2; 1; 2; KOH. $2KOH + H_2SO_4 \rightarrow H_2O + K_2SO_4$

117. Addition of the coefficient 2 before KOH has balanced the equation for K. The equation is not balanced for H or for O. For example, there are six oxygens on the reactant side (oxygen is present in both reactants) and only five on the product side. All elements must be present in equal amounts on both sides of the equation to achieve balance. Four hydrogens are present in the reactants (two are present in the H_2SO_4 molecule; there is one hydrogen in each of the two KOH molecules as well); 2; 2.

118. $2KOH + H_2SO_4 \rightarrow 2H_2O + K_2SO_4$ The equation is balanced.

119. H^+ and NO_3^-.

120. releases OH^- and NH_4^+ ions

121. neutralization; H_2O and NH_4NO_3

The salt ammonium nitrate forms when the ammonium ion, with a charge of +1, combines with the nitrate group, with its charge of –1. Ammonium and nitrate are both polyatomic ions.

122. $HNO_3 + NH_4OH \rightarrow H_2O + NH_4NO_3$

123. H_2O and $(NH_4)_2SO_4$.

This salt, ammonium sulfate, combines the positively charged ammonium ion with the sulfate ion. Because the sulfate group has a charge of –2 and the ammonium group has a charge of +1, it is necessary to combine two ammonium groups with one sulfate group to form a neutral compound. The subscript 2 shows that two ammonium groups are present for each sulfate.

124. The problem can be started by listing the reactants on the left and the products on the right. Counting the number and kinds of atoms on each side of the equation shows that the equation is not balanced. For example, there is one nitrogen on the reactant side and two nitrogens on the product side of the equation.

$$H_2SO_4 + NH_4OH \rightarrow H_2O + (NH_4)_2SO_4.$$

The equation can be balanced for nitrogen by adding a coefficient of 2 in front of the ammonium hydroxide molecule. Remember, there are now two copies of the entire molecule present, and all the atoms in it are present in doubled amounts. The equation now has the following form:

$$H_2SO_4 + 2NH_4OH \rightarrow H_2O + (NH_4)_2SO_4.$$

A tally of the atoms shows that there are now twelve hydrogens on the left and ten on the right. There are six oxygens on the left and five on the right. These elements can be balanced by placing a coefficient of 2 in front of the water molecule. The following equation is balanced.

$$H_2SO_4 + 2NH_4OH \rightarrow 2H_2O + (NH_4)_2SO_4.$$

125. neutralization; 7.

126. decrease; less.

127. resists or minimizes.

128. hydrogen; ionic; independently of each other; electrolyte.

129. A double arrow is used to show that a reaction is reversible.

The reaction does not go to completion. At any time, there is some carbonic acid in the dissociated state and much that has not dissociated. All the components shown in the equation are present at the same time. The molecules are constantly reacting in both the forward and reverse directions.

130. dissociate and move separately; remain together as a unit. The bond between hydrogen and the bicarbonate group (HCO_3^-) is strongly polar. Such a molecule has some ionic character. It is an intermediate case, neither fully ionic nor fully covalent. Partial dissociation occurs in this situation.

131. hydroxide; H^+; accept; reversible. At completion (or equilibrium) for the above reaction, there is a mixture of carbonic acid, hydrogen ions, and bicarbonate.

132. weak acid; weak base; buffer.

133. acid; hydrogen; hydrochloric acid; bicarbonate (HCO_3^-); carbonic acid in the buffer system and less bicarbonate (some of it was used up in the reaction and converted to carbonic acid); less.

134. strong base; hydroxide ions; hydroxide; H^+. more bicarbonate and less carbonic acid (some of the carbonic acid is used up in the reaction with the incoming base); less.

135. the reaction is reversible, and both reactants and products are present. carbonic acid; does not.

136. carbonic acid; hydrogen ions; bicarbonate ions.

137. Cells generate carbon dioxide as they live and metabolize. Because water is the solvent of the cytoplasm, it is always available. As cells produce carbon dioxide, it dissolves in water. Some of the carbon dioxide then reacts with the water to produce carbonic acid. Some carbonic acid goes through the reversible dissociation to produce hydrogen ions and bicarbonate groups. Thus, the normal life of the cell generates the compounds for this buffered system. The sequence of two reactions is shown in the two connected reactions of the equations below.

$$CO_2 + H_2O \rightleftarrows H_2CO_3 \rightleftarrows H^+ + HCO_3^-$$

138. bicarbonate.

139. 1 amu or dalton; atomic mass units (amu) or daltons; 2 (hydrogens); 2 daltons (weight); oxygen; 16 daltons (weight); 3 (atoms); 18 daltons (weight of a water molecule); 18 amu or daltons.

140.

Atom	Number in Molecule	Atomic Weight	Total Weight
C	6	12	72
H	12	1	12
O	6	16	96
		Molecular Weight =	180

The atomic weight for an element is multiplied by the number of atoms of that element. This calculates the total weight contribution from that element. For example, six atoms of carbon × 12 daltons per carbon = 72 daltons for the total contribution from carbon. At the end, the weights from all the atoms are added to obtain the weight for the entire molecule.

141. $(NH_4)_3PO_4$.

Atom	Number in Molecule	Atomic Weight	Total Weight
N	3	14	42
H	12	1	12
P	1	31	31
O	4	16	64
		Molecular Weight =	149

Note that there are three copies of the ammonium group, NH_4, in the ammonium phosphate molecule. Everything in the ammonium group is therefore multiplied by 3 to calculate the total number of atoms present in the molecule. The entire molecule has a molecular weight of 149 daltons.

142. 18 amu or daltons; 18; grams (for water); 180 (daltons); 180 grams (of glucose).

143. Molecular weight is the weight of one molecule. It is calculated by adding the weights of all the atoms that are present in one molecule. The units are atomic mass units (amu) or daltons. A mole is a gram molecular weight, that is, the molecular weight expressed in grams. The molecular weight is the weight of one molecule in daltons. All moles contain the same number of molecules, Avogadro's number, equal to 6.022×10^{23}. The reason all moles have the same number of molecules is that the value of the mole is proportional to the molecular weight.

144. 18 daltons (one molecule of water); 180 daltons (one molecule of glucose); 10 times; 1 (molecule of each); equal.

145. 180 daltons; 1800 daltons; equal; 10 times.

146. 1800 daltons; 18,000 daltons; equal; 10 times.

147. 10×; 10 times.

148. 18 grams; 180 grams; the same as.

Because the mass of the glucose is ten times greater than the mass of the water, the number of molecules will be equal. One glucose weighs ten times more than one water. As long as this proportion of 10:1 is maintained, the number of molecules will be equal. Individual molecules are very tiny, so the number of molecules in a mole is extremely large. This value is Avogadro's number and equals 6.022×10^{23}.

149. 6.022×10^{23}; 6.022×10^{23} (molecules).

150. 1.36 moles of glucose.

151. 0.72 moles of ammonium phosphate.

152. moles; liter.

153. 0.46M; an equal (concentration of solute molecules); 10 ml.

154. 180 (grams); 180 (grams); 1M; 1M $C_6H_{12}O_6$ (glucose).

155. The molecular weight of sucrose is calculated by adding the weights of all the atoms that are present in one molecule. C = 12 × 12 = 144 daltons. H = 1 × 22 = 22 daltons. O = 16 × 11 = 176 daltons. The sum of the weights contributed by carbon, hydrogen, and oxygen is the molecular weight: 144 daltons (contributed by 12 carbons) + 22 daltons (contributed 22 hydrogens) + 176 daltons (contributed by 11 oxygens) = 342 daltons = the molecular weight of sucrose. Weigh 342 grams of sucrose (this is one mole of sucrose). Dissolve the sucrose in distilled water. When the sucrose is fully dissolved, add water sufficient to make exactly 1 liter of solution. Label the solution.

156. 180 grams; 6.022×10^{23} molecules; 342 grams; 6.022×10^{23} molecules; equal.

157. 180 grams; 1 mole; 342 grams; 1 mole; equal numbers of solute molecules; equal volumes of solution; equal concentrations; 3 ml.

158. moles per liter; moles.

Moles/liter × liter = moles. The volume (in metric units of the liter) in the denominator of the first term is cancelled by the volume in the numerator of the second term.

159.*The volume and the concentration are multiplied to obtain the answer (formula 1 is used first). 0.50M × 500 ml = ? To solve the problem, replace molarity (M) by moles per liter, and convert 500 ml to 0.5 liter. The problem now reads: 0.50 moles/liter × 0.5 liter = ? Multiplying the two values gives the answer of 0.25 moles. The volume units have cancelled. The amount of glucose required for preparation of this solution is 0.25 moles.

160. grams per mole; grams.

161.*Formula 2 is used to solve this problem. The number of moles needed was calculated using formula 1 in question 159. The molecular weight of glucose was calculated previously. Substituting these values in formula 2 gives:

180 grams/mole of glucose × 0.25 moles of glucose = 45 grams of glucose

Note that moles in the denominator of the first term cancels moles in the second term. The answer has units of grams.

162.*45 (grams of glucose to weigh); distilled water; 500 ml; 0.50M glucose.

163. 0.50M. The concentration is the same whether the sample is a large or a small volume. The solution is homogeneous or the same throughout.

164.** Formula 1 is used first: 0.36 moles/liter × 0.4 liter = 0.144 moles
Formula 2 is used second: 180 grams/mole × 0.144 moles = 25.92 grams

165.*Weigh out 25.92 grams of glucose, and dissolve it in pure distilled water. When the glucose is fully dissolved, add distilled water to exactly 400 ml. Store and label as 0.36M glucose.

166. $C_3H_7O_3N$ formula for serine.

Atom	Number in Molecule	Atomic Weight	Total Weight
C	3	12	36
H	7	1	7
O	3	16	48
N	1	14	14
		Molecular Weight =	105

167. Formula 1: 0.8 moles/liter × 0.650 liters = 0.52 moles
Formula 2: 105 grams/mole × 0.52 moles = 54.6 grams

168. The calculations in the previous problem show that 54.6 grams of serine is required. Carefully weigh out 54.6 grams of serine and dissolve it in distilled water. After it is dissolved, add distilled water to exactly 0.650 liters (650 ml). Label the solution as 0.8M serine.

169. 100 ml of a 1.5M solution of glucose.
Formula 1: 1.5 moles/liter × 0.1 liter = 0.15 moles
Formula 2: 180 grams/mole × 0.15mole = 27 grams

Dissolve 27 grams of glucose in distilled water and bring to a final volume of 100 ml.

170. 750 ml of a 0.2M solution of sucrose.
Formula 1: 0.2 moles/liter × 0.750 liter = 0.15 moles
Formula 2: 342 grams/mole × 0.15 moles = 51.3 grams

Dissolve 51.3 grams of sucrose in distilled water and bring to a final volume of 750 ml.

171. 95 ml of a 0.04M solution of serine.

Formula 1: 0.04 moles/liter × 0.095 liter = 0.0038 moles

Formula 2: 105 grams/mole × 0.0038 moles = 0.399 grams

Dissolve 0.399 grams of serine in distilled water and bring to a final volume of 95 ml.

172. 4M; 2M; 200 ml.

173. Volume of stock solution:

$$4M \times IV = 2M \times 200\text{ ml}; \quad \text{rearrange} \rightarrow IV = \frac{2M \times 200\text{ml}}{4M} \text{ and } IV = 100\text{ ml}$$

How much water?

Final volume – Initial volume = Amount of water needed. 200 ml – 100 ml = 100 ml. Take 100 ml of the 4M solution and add 100 ml of distilled water. Mix thoroughly. There will now be 200 ml of a 2M NaCl solution.

174. IC = 3M, FV = 500 ml, and FC = 0.25M. Substitute these values in the formula and:

$$IV = \frac{0.25M \times 500\text{ml}}{3M} \text{ ml} \quad \text{and} \quad IV = 41.67\text{ ml}$$

Measure out 41.67 ml of the 3M stock solution of glucose. Add water up to 500 ml and mix thoroughly.

175. IC = 1M, IV = 20 ml, and FC = 0.015M. Substitute these values into the formula and rearrange.

$$\frac{1M \times 20\text{ ml}}{0.015M} = FV \text{ and } FV = 1333.33\text{ ml} = 1.33333\text{ liters}$$

Use all 20 ml of the 1M stock solution. Add distilled water to the 20 ml of stock solution to a final volume of 1333.33 ml of 0.015M alanine. This is the maximum amount that can be prepared from the remaining stock solution.

176. When a concentration is multiplied by a volume, the result is a certain number of moles. (See formula 1 in the previous sections.) This calculation yields the number of moles of solute present in a solution. After a solution is diluted, the same amount of solute is present. There is now a larger volume of a more dilute solution. The same amount of solute is present regardless of the amount of water diluting it.

Chapter Test: Aqueous Chemistry

The questions in this chapter test evaluate your mastery of all the objectives for this unit. Material from earlier chapters is not tested directly, but recall of earlier material is necessary background in some cases. Take this test without looking up material in the chapter. You may use the periodic table. Questions that test intermediate or advanced objectives are indicated by asterisks, and you may omit them if you did not cover those topics in the chapter. After you have completed the test, you can check your work by using the answer key located at the end of the test.

1. The compound that is present in largest amount in living organisms is ___________.

2. The shape of molecules such as methane, ammonia, and water is

   a. linear. b. planar. c. square. d. tetrahedral. e. none of these.

2. Methane is a

   Polar______ Nonpolar ______ molecule. Explain your answer.

4. In the space below, sketch a diagram of the water molecule. Indicate the location of partial charges. Explain why water molecules have this shape, and explain why they are polar.

5. Which statement is not true of hydrogen bonds? Hydrogen bonds
   a. are often weak attractive forces between different molecules.
   b. can form only when a molecule contains hydrogen bonded to fluorine, oxygen, or nitrogen.
   c. connect atoms together into molecules.
   d. can form between different water molecules, causing them to exhibit chemical stickiness.
   e. give water some unique properties, such as its high heat capacity.

6. Cytoplasm is a complex aqueous solution. The solvent is ___________, and the solutes include ___________, ___________, and ___________. (Give at least three examples.)

7. List the classes of compounds that usually dissolve in water.

What kinds of compounds usually do not dissolve in water?

8.*Predict whether each of the compounds listed below is polar or nonpolar. Then predict whether it will dissolve in water. Explain the reasons for your predictions.

Compound	Polar or Nonpolar Molecule	Water Solubility
Ethane—C_2H_6		
Alanine—$C_2H_7O_2N$		
$MgCl_2$		
NH_3		

9. Solutes that release ions in solution are called ________________________, because they ____________________________. This permits the solution to conduct an electrical current.

10. $CaCl_2$ is the formula for __________________, which is a(n) ____________ and also a salt. In the space below, sketch the particle(s) that $CaCl_2$ releases in aqueous solution. Show how they would be hydrated.

11. A strong base is a compound that releases ______. An example of such a base is ____________. A more general definition of a base is that it is a substance that can __________________. An example of a weak base that fits the second definition but not the first is ____________.

12. H_3PO_4 is called ___________. It is classified as a solute that is an ______ and ___________. In aqueous solution, this molecule releases ______ (what number) ions. List the ions that are released by this compound. ___________. The pH of a solution would ___________ after this compound is dissolved in it.

13. An acid would have a pH below ___________.

14. HNO_3 is a compound called _________________. $Ca(OH)_2$ is a compound called ___________. When these two compounds are mixed, a ___________ reaction occurs.

15. What ions form when HNO_3 is dissolved in water? ______ and ______ What ions form when $Ca(OH)_2$ is dissolved in water? ______ and ______ What products will form after the reaction of these two compounds has occurred? ______ and ______

16.*Write and balance the equation for the reaction in the previous problem. Show your work.

17. A substance that resists a change in pH is a(n)
   a. acid. b. base. c. buffer. d. electrolyte. e. polar molecule.

18. Calculate the molecular weight of ethanol, C_2H_6O. Show your work.

19. A mole is defined as a ________________________________.

20. A mole of alanine, a mole of glucose, and a mole of water all have ___________ numbers of molecules.

21. What is Avogadro's number? ____

22.**List the two formulas that are used to solve molarity problems.
Formula 1 __________________ Formula 2 __________________

23.**How many moles of ethanol are required to prepare 400 ml of a 0.20M solution?

24.**How many grams of ethanol are required for the solution requested in question 23? ______

25. Explain how to prepare the solution in problem 23.

26. Explain how you would prepare 200 ml of a 0.05M solution from a 2M stock solution.

Answers

1. Water.

2. d. tetrahedral.

3. Nonpolar molecule.

   Methane contains four hydrogens linked to a central carbon by slightly polar covalent bonds. The molecule has a tetrahedral shape. The slight displacements of the electron cloud in the polar covalent bonds offset each other, because they pull in opposing directions. The molecule as a whole is therefore nonpolar.

4.

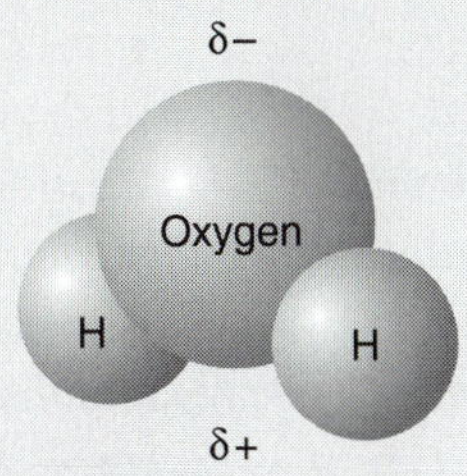

δ−

Oxygen

H H

δ+

   Water molecules have the shape of a tetrahedron. Oxygen is at the center. There are two hydrogens bonded to oxygen that are located at two of the apexes of the regular tetrahedron. The other two apexes are occupied by unbonded electron pairs that are present in the valence electron shell of oxygen. The oxygen end of the water molecule develops a partial negative charge because oxygen is an electronegative element that attracts the bonding electrons toward itself. Also, there are two unbonded pairs of electrons in this region. The slightly positive pole develops at the hydrogen end of the water molecule, because oxygen shifts the electron cloud away from this region.

5. c. connect atoms together into molecules. This statement is the only answer that is false.

6. water; many electrolytes such as NaCl, many small polar organic molecules such as sugars, and dissolved gases (CO_2 and O_2 are valid examples here).

7. Molecules that are polar and compounds containing ions are often soluble. Nonpolar molecules such as the hydrocarbons are normally not water soluble.

8.*

Compound	Polar or Nonpolar Molecule	Water Solubility
Ethane—C_2H_6	Nonpolar	Insoluble
Alanine—$C_2H_7O_2N$	Polar	Soluble
$MgCl_2$	Ionic	Soluble
NH_3	Polar	Soluble

Ethane contains only hydrogen and carbon. Hydrocarbons are nonpolar and do not dissolve in water.

Alanine contains three atoms (two oxygens and one nitrogen) that are very electronegative. These elements attract the electron cloud and create poles of partial charge in the molecule. This polar molecule is water soluble.

Magnesium chloride is a salt. It contains ions that dissociate and move independently into the aqueous solution. Many salts, including magnesium chloride, are water soluble.

Ammonia, or NH_3, is polar for the same basic reasons that water itself is polar. The ammonia molecule has tetrahedral symmetry. One apex is occupied by a nonbonding pair of electrons. The other three apexes are occupied by hydrogens. The end of the molecule with the nonbonding pair of electrons and the nitrogen is electronegative. Ammonia is, therefore, polar and because of that, is highly water soluble.

9. electrolytes; release mobile charged particles in solution.

10. calcium chloride; electrolyte.

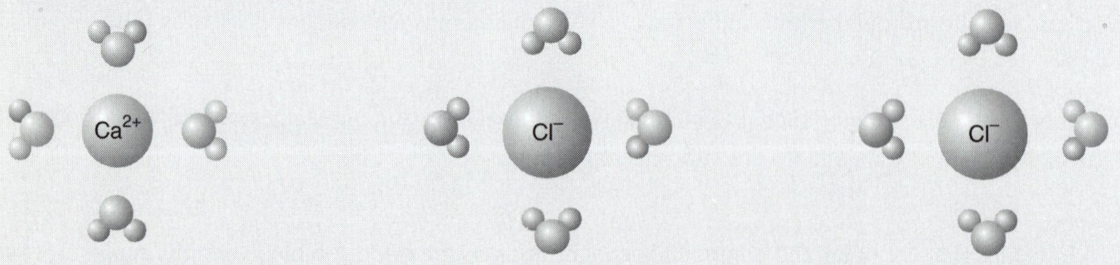

Ca^{2+} Cl^- Cl^-

11. OH^-; NaOH, KOH, $Ca(OH)_2$, or NH_4OH are all valid answers; accept hydrogen ions; bicarbonate, HCO_3^-.

12. phosphoric acid; electrolyte and acid; four ions; three H^+ ions and one phosphate group—PO_4^{3-}; decrease.

13. 7.

14. nitric acid; calcium hydroxide; neutralization.

15. H^+ and NO_3^-; Ca^{2+} and $2OH^-$; H_2O and $Ca(NO_3)_2$

16.* The first draft of the equation lists the reactants on the left side of the equation and the products on the right.

$$HNO_3 + Ca(OH)_2 \rightarrow H_2O + Ca(NO_3)_2$$

Analysis of this equation shows that it is not balanced. For example, there is one nitrogen on the left, and two on the right (remember that everything in parentheses is doubled). Addition of a coefficient of 2 in front of the nitric acid will achieve balance for nitrogen.

$$2HNO_3 + Ca(OH)_2 \rightarrow H_2O + Ca(NO_3)_2$$

A check of hydrogen shows that there are four hydrogens on the reactant side (left) and only two on the product side. This can be remedied by placing a coefficient of 2 in front of water.

$$2HNO_3 + Ca(OH)_2 \rightarrow 2H_2O + Ca(NO_3)_2$$

The equation is now balanced. The numbers of all types of atoms are the same on both sides of the equation.

17. c. buffer.

18.

Carbon weighs 12×2 atoms per molecule	=	24 daltons.
Hydrogen weighs 1×6 atoms per molecule	=	6 daltons.
Oxygen weighs 16×1 atom per molecule	=	16 daltons.
Molecular weight	=	46 daltons.

19. gram molecular weight.

20. equal.

21. Avogadro's number is 6.022×10^{23}. This is the number of molecules in a mole.

22.** Formula 1 Molarity × Volume = Number of Moles
Formula 2 Number of Moles × Mass of 1 Mole = Grams

23.**400 ml of 0.20M solution:

$$0.20 \text{ moles/liter} \times 0.4 \text{ liters} = 0.08 \text{ moles}$$

24.**0.08 moles × 46 grams/mole = 3.68 grams (grams of ethanol required)

25. Weigh out 3.68 grams of ethanol and add water until the final volume is 400 ml.

26. 200 ml of a 0.05M solution from a 2M stock solution:
The formula IC × IV = FC × FV can be used here.

$$IV = \frac{0.05M \times 200\,ml}{2M} = 5.0 \text{ ml}$$

Carefully measure 5.0 ml of the 2M stock solution. Add water up to a final volume of 200 ml and mix.

Chapter 5

Organic Chemistry

Objectives

1. Define the following terms and apply the definitions correctly: alcohol, aldehyde, amine, carbonyl, carboxyl, functional group, isomers, ketone, organic chemistry, phosphate, sulfhydryl, and thiol.
2. List the properties of carbon that are responsible for the great variety of bonding patterns observed in organic compounds.
3. Identify hydrocarbons and list the properties of this group of compounds. Interpret structural formulas, condensed structures, and line-structure representations of hydrocarbons.
4. Identify alcohols and list their properties.
5. Identify thiols and list their properties.
6. Identify carbonyl groups within organic compounds. State the similarities and differences between aldehydes and ketones. List the properties of these groups.
7. Identify carboxyl groups and carboxylic acids. Explain why carboxylic acids exhibit acidic properties.
8. Identify amines and explain how they can react as bases.
9. Identify phosphates and list their properties.

The Central Role of Carbon

All organic compounds are based on the element carbon. Organic compounds form most of the dry weight of all living organisms. Organic compounds vary greatly in size and structure and are therefore well suited to provide the complexity required in living systems. The great variety of structures found in organic chemistry occurs because carbon forms four covalent bonds, can bond to itself and to many other elements as well, and forms carbon backbones with a great variety of bonding patterns.

Questions

1. Organic compounds always contain the element ______.

2. Suppose a jellyfish is dehydrated in a slow drying process. After dehydration, the jellyfish would weigh ______ (more or less) than before, due to the loss of ______. The remaining weight would be the dry mass of the jellyfish and would consist mostly of ________________________.

3. There are more kinds of organic than inorganic compounds. This means that there are more compounds based on the element ______ than there are compounds of all the other elements.

4. Carbon has ______ valence electrons and tends to ______ four more electrons to create an octet of electrons in the valence shell. Carbon forms ______ covalent bonds to achieve stability.

5. Two carbon atoms can share one pair of electrons and form a ______ bond, or share ____________ electrons to form a double bond, or share three pairs of electrons to form a ______ bond. Bonds between two carbon atoms are always ____________, because all carbon atoms have the same electronegativity.

6. Hydrogen has ______ electron(s) and can bond to carbon by forming a ____________. Hydrogen forms ______ bond(s) when it shares electrons.

7. Look up the electronegativities of each of the elements listed in the table below on page 98. Then predict what type of bond would form when the element reacts with carbon, and explain why this type of bond forms.

Element	Electronegativity	Bond Formed with Carbon
C		
H		
O		
N		
S		

Organic compounds have complex and varied structures that result from the properties of carbon. Although there are enormous numbers of organic compounds, systematic study shows that certain patterns of atoms and bonds recur. These are called the organic functional groups, and each has predictable properties. This chapter presents the organic functional groups that are found in biologically important compounds. Familiarity with the organic functional groups presented in this chapter will prepare you for first year college biology. An organic chemistry course will include additional functional groups, an extensive treatment of reaction mechanisms, and a more complete analysis of molecular shape.

Questions

8. It is possible to systematize the study of organic compounds and predict their properties, because there are patterns of atoms and bonds that are found in many different organic compounds. These patterns are called the ________________, and the properties of a particular group are predictable.

9. Explain why organic compounds are so varied and complex. ______________________________

10. When two atoms that have the same electronegativity react, they share electrons ______ and form ____________. When two atoms share electrons but differ noticeably in electronegativity, they form ________________.

11. Carbon has an electronegativity that is intermediate in value. This means it can form nonpolar covalent bonds with some elements and ____________ bonds with many other elements, such as oxygen and nitrogen.

12. Organic chemistry is complex and varied, because carbon has such a versatile bonding pattern. Carbon exhibits this variety, because it can form ______ bonds, has an ____________ electronegativity and can therefore bond to ______ other elements.

13. Compounds held together by ionic bonds usually dissociate and release ______ when they dissolve in water. The atoms in organic compounds are held together by ____________. Some organic compounds are water soluble, and when they dissolve in water, they ______ (do or do not) dissociate and ______ release ions, and they are nonelectrolytes.

14. The organic compounds that contain only ______ and ______ are called hydrocarbons. The formula for methane is ______, and the formula for ethane is ______ (see page 131). These compounds contain only carbon and hydrogen and are therefore ______. Methane and ethane ______ dissolve in water, because they are ______ (polar or nonpolar) molecules.

15. Compounds that do dissolve in water are hydrophilic, and those that do not are ____________. Methane and ethane are ____________. Would you predict that other hydrocarbons would be water soluble? Explain your answer.

Hydrocarbons

The diagrams below show the structures of several compounds that are classified as hydrocarbons. They are organic compounds that contain only the elements hydrogen and carbon. Hydrocarbons are nonpolar molecules and are therefore hydrophobic.

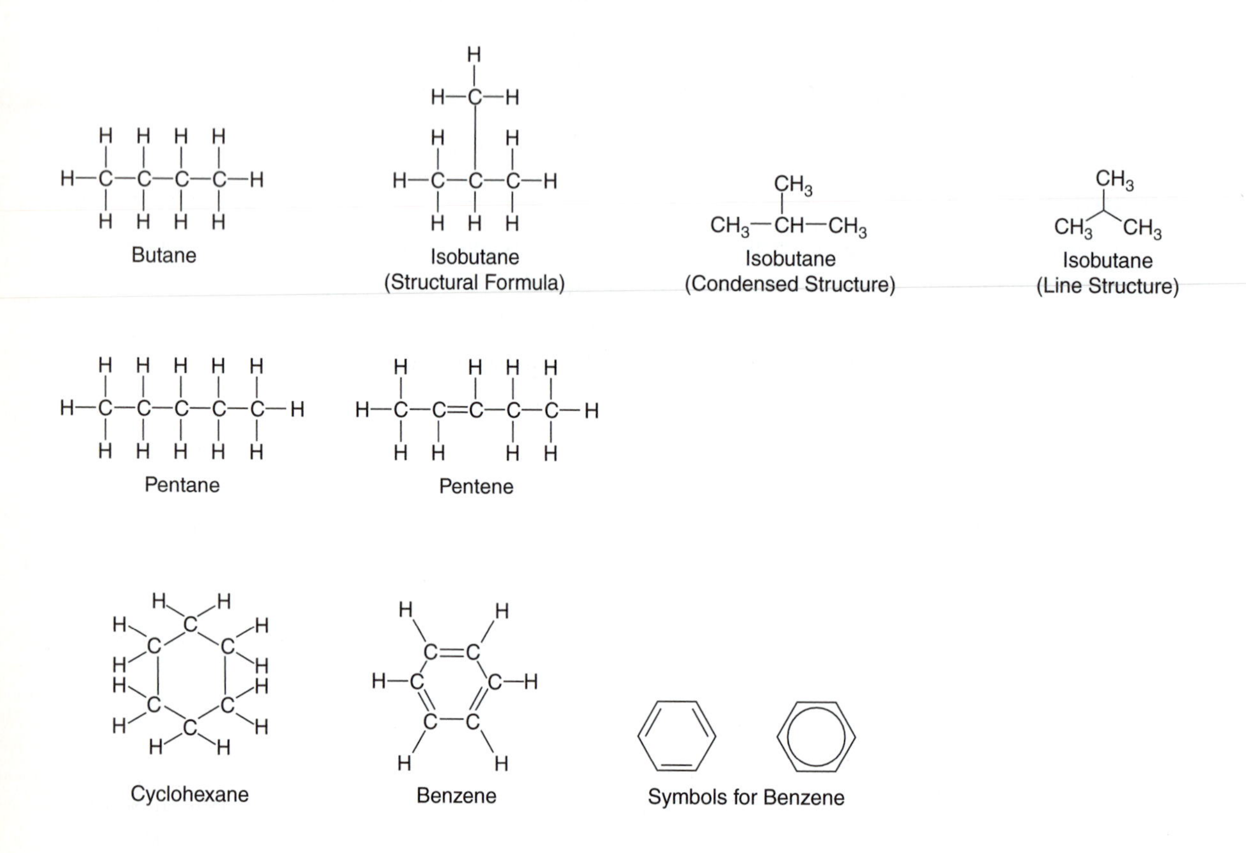

Questions

Answer all the questions by referring to the structures of the hydrocarbons in the diagram above.

16. Which of the molecules above have backbones of carbon that do not branch? ____________________.
    Which structures do contain branches in the carbon chain? ____________________.

17. Explain how the structures of cyclohexane and benzene differ from the other structures.

18. How does pentene differ from pentane? ____

19. Structures containing carbon are very diverse. Based on your answers to the last three questions, summarize the reasons for this.

20. The molecular formula for butane is ____________, and the molecular formula for isobutane is ____________. The bonding pattern and structural formulas of butane and isobutane ____________ the same. Two compounds that have the same molecular formula and different structural formulas are called isomers. Butane and isobutane ______ (are or are not) isomers and have similar chemical properties.

21. The structure of butane can be represented in an abbreviated way as $CH_3CH_2CH_2CH_3$. The CH_3 group is called a methyl group. Circle the methyl groups on the ends of the butane molecule in the diagram to the left. Carbons that are not located on the ends of the butane molecule are bonded to just ______ hydrogens instead of three. Explain why this is so.

22. Methane is not soluble in water because it is ____________. Other hydrocarbons are similar and do not dissolve in water; these are referred to as ____________.

23. Methane is the major component in natural gas. When natural gas is burned (oxidized), it ____________ (releases or absorbs) a great deal of energy. Hydrocarbons can therefore be used as ____________.

24. Fossil fuels such as coal and petroleum can be burned with the release of a great deal of ____________. Fossil fuels are rich in ____________ derived from organic molecules that are present in the remains of organisms that lived long ago.

25. If a liquid that is rich in hydrocarbons (such as gasoline) is mixed with water, it ______ (does or does not) dissolve, because hydrocarbons are ____________ and therefore ____________. Because they contain so much hydrogen, hydrocarbons have a low density. The gasoline would therefore tend to form a layer ____________ the water.

26. How does the structure of benzene differ from that of cyclohexane?

27. Structures such as benzene that have alternating double and single bonds are flat or planar in shape. The electrons actually form a cloud above and below the flat ring and are described as delocalized, because the cloud is spread evenly over the entire molecule. Compare the structure of cyclohexane to that of benzene. How are they the same? ______________________ How do they differ? ______________________ Would you predict that cyclohexane would be a planar molecule, and why do you make this prediction?** ____________________________

28. Flat rings with delocalized electrons exhibit some unique properties (such as absorption of ultraviolet radiation) and are referred to as aromatic rings. Benzene ______ an aromatic ring. The benzene ring can be represented with symbols that are simpler than the full structure. What are these symbols?

29. Benzene rings are not found as independent molecules in living organisms, but rings of this type are found as components of larger molecules, such as some amino acids. Amino acids that contain a benzene ring within their structures ___________ aromatic and would absorb ________________.

30. Organic chemists often use a kind of shorthand to represent the structures of compounds. Compare the condensed structure of isobutane to the full structural formula. All of the atoms that are present in isobutane ___________ (are or are not) listed in the condensed structure. All of the covalent bonds that are present in the structural formula ___________ (are or are not) present in the condensed structure.

31. Compare the line structure of isobutane to the condensed structure and structural formulas. All of the atoms that are present in isobutane ___________ (are or are not) listed in the line structure. All of the covalent bonds that are present in the structural formula ___________ (are or are not) listed in the line structure.

32. Compare the symbols for benzene to the complete structural formula for benzene. The carbon atoms that form the benzene ring ___________ (are or are not) shown in the symbols for benzene. The six hydrogen atoms that are present in benzene ___________ shown in the benzene symbols and are understood to be present.

33. Sometimes hydrogens are omitted from shorthand formulas and symbols for compounds. How can you determine how many hydrogens are actually present, even though they are not shown?

34. Sometimes carbons are omitted from shorthand formulas and symbols for compounds. How can you determine where carbons are located, even though they are not shown directly?

35. Hydrocarbons as a group ____________ (do or do not) tend to dissolve in water and are therefore __________________. Hydrocarbons have a low density and, when mixed with water, tend to ____________. The energy content of hydrocarbons is ____________.

36. One group of biochemicals found in cells is made mostly of hydrocarbons. Which of the following groups has properties most similar to those summarized in the previous question? a. carbohydrates such as sugar and starch b. lipids such as fats and oils c. proteins and amino acids (egg white is rich in a soluble protein)

Alcohols

One or more of the hydrogens in a hydrocarbon can be replaced by another functional group. The properties of the hydrocarbon are often changed drastically by the newly introduced functional group. Replacement of a hydrogen by a hydroxyl group (–OH) forms an organic compound classed as an alcohol. The pattern of atoms and bonds that is characteristic of an alcohol is shown below (*a*). The structures of some representative alcohols are shown in *b*.

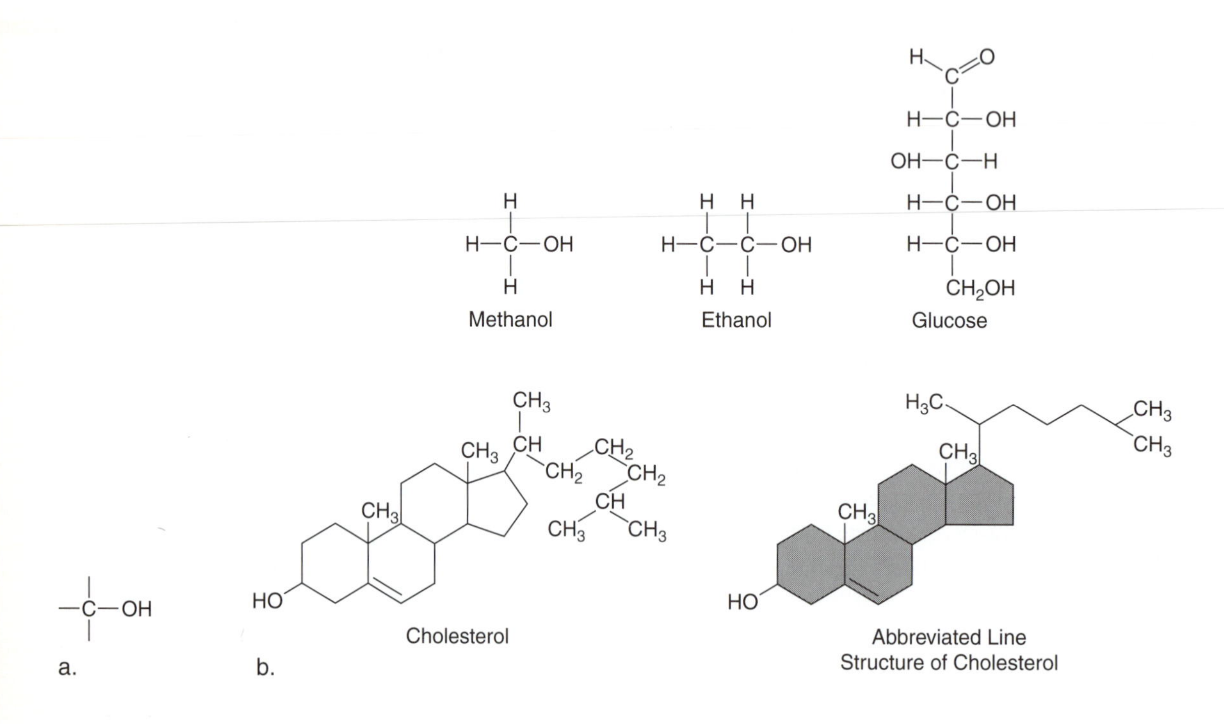

Questions

37. Bonds between carbon and oxygen are ____________ (polar or nonpolar) covalent bonds. Replacing a hydrogen with a –OH group will tend to make an organic compound ____________ (more or less) polar.

38. Compounds that are nonpolar do not dissolve readily in water and are said to be ____________. Compounds that are ____________ (polar or nonpolar) usually are soluble in water and are ____________.

39.* Compounds that can form hydrogen bonds are even more likely to be water soluble than compounds that are simply polar. Explain why this is so.

40. Hydrogen bonds are ____________ (weak or strong) bonds that ____________ (do or do not) link atoms into molecules. What circumstances lead to the formation of hydrogen bonds? (Review page 135 if you do not remember this.)

41. Alcohols ____________ (can or cannot) form hydrogen bonds. The presence of one or more hydroxyl groups in an organic compound tends to ____________ its solubility in water.

42. Compare the structure of an alcohol and a hydrocarbon of the same size, such as ethanol and ethane. Predict which would be more soluble in water and explain your reasons.

43. The bond between carbon and oxygen is ____________, and ____________ dissociate when the compound dissolves in water. Predict whether alcohols are electrolytes or nonelectrolytes and explain your reasoning.

44. Many alcohols have names that end in ____________, indicating that they contain ____________ functional group.

45. Glucose is a sugar that is central in the metabolism of most organisms. Glucose contains ____________ (how many) alcohol groups. Predict whether glucose will dissolve in water, and explain your reasoning.

46. Glycerol is one of the building blocks used to make some lipids (see page 257). Glycerol contains three ____________ groups and could also be called a trialcohol.

47. Circle the alcohol group in the cholesterol molecule. All of the cholesterol molecule except for this alcohol group is composed of ____________. The alcohol group is polar and can also form __________________, factors that tend to promote solubility in water. Hydrocarbons are nonpolar and ____________ soluble in water. The majority of the cholesterol molecule is composed of __________________.

48. Predict whether cholesterol will dissolve in water, and explain the reasons for your answer.

49. Compare the two representations of the cholesterol molecule. Both show a set of connected rings. These rings are made of ____________. In addition to the bonds to other carbons, the atoms in the ring are also attached to enough ____________ to complete their valence requirement of four bonds.

50. The set of connected rings in the cholesterol molecule is found in all compounds classed as steroids. The rings are made of __________________ and tend to be nonpolar; therefore, steroids have little tendency to __________________.

51. Explain how the abbreviated structure of cholesterol differs from the more complete structure.

Thiols

Sulfur has six valence electrons, is in the same chemical group as oxygen, and forms two covalent bonds (as does oxygen). The sulfhydryl group (–SH) is the sulfur-containing analog of the hydroxyl group (–OH). Organic compounds that contain a sulfhydryl group linked to carbon are called thiols. The structures of some thiols are shown in the diagram below.

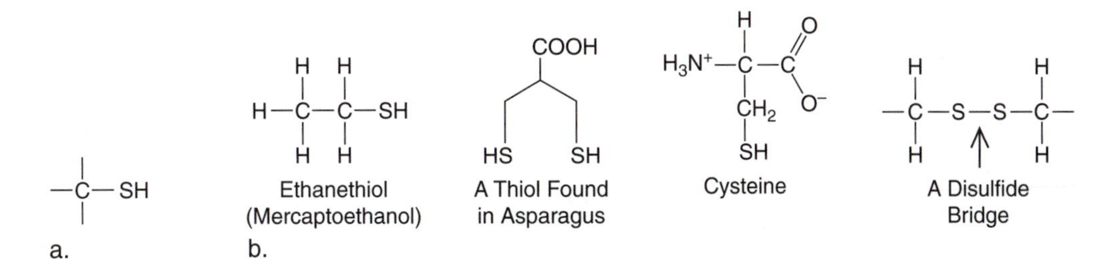

Questions

52. Sulfur has six valence electrons and can achieve stability by sharing ______ (how many) electrons to create an ____________. Sulfur usually forms ______ (how many) covalent bonds. Oxygen also forms ______ covalent bonds.

53. The thiol found in asparagus contributes to the unique smell of that vegetable. Thiols are also found in coffee, onions, and garlic. Hydrogen sulfide (H_2S) smells like rotten eggs. Speculate as to how the presence of sulfur in a compound is related to its smell.

54. Although thiols are similar in some respects to alcohols, they also differ, mainly because sulfur is ______ electronegative than oxygen. Alcohols ______ form hydrogen bonds. Predict whether thiols can form hydrogen bonds, and explain the reasons for your prediction.

55. Two amino acids contain sulfur. The amino acid in the diagram above is ______. It contains a sulfhydryl group and is therefore a ____________.

56. When two cysteine molecules are near each other, they can react and form a disulfide linkage, a covalent bond that links them together through their sulfur atoms. Covalent bonds are ____________ (strong or weak) bonds that ____________ dissociate in water. Reactions of this type play an important role in establishing and maintaining the shape of many proteins.

Aldehydes and Ketones

An oxygen linked to a carbon with a double bond is called a carbonyl. Aldehydes and ketones are two organic functional groups that contain a carbonyl. Examples of aldehydes and ketones are shown in the diagram below.

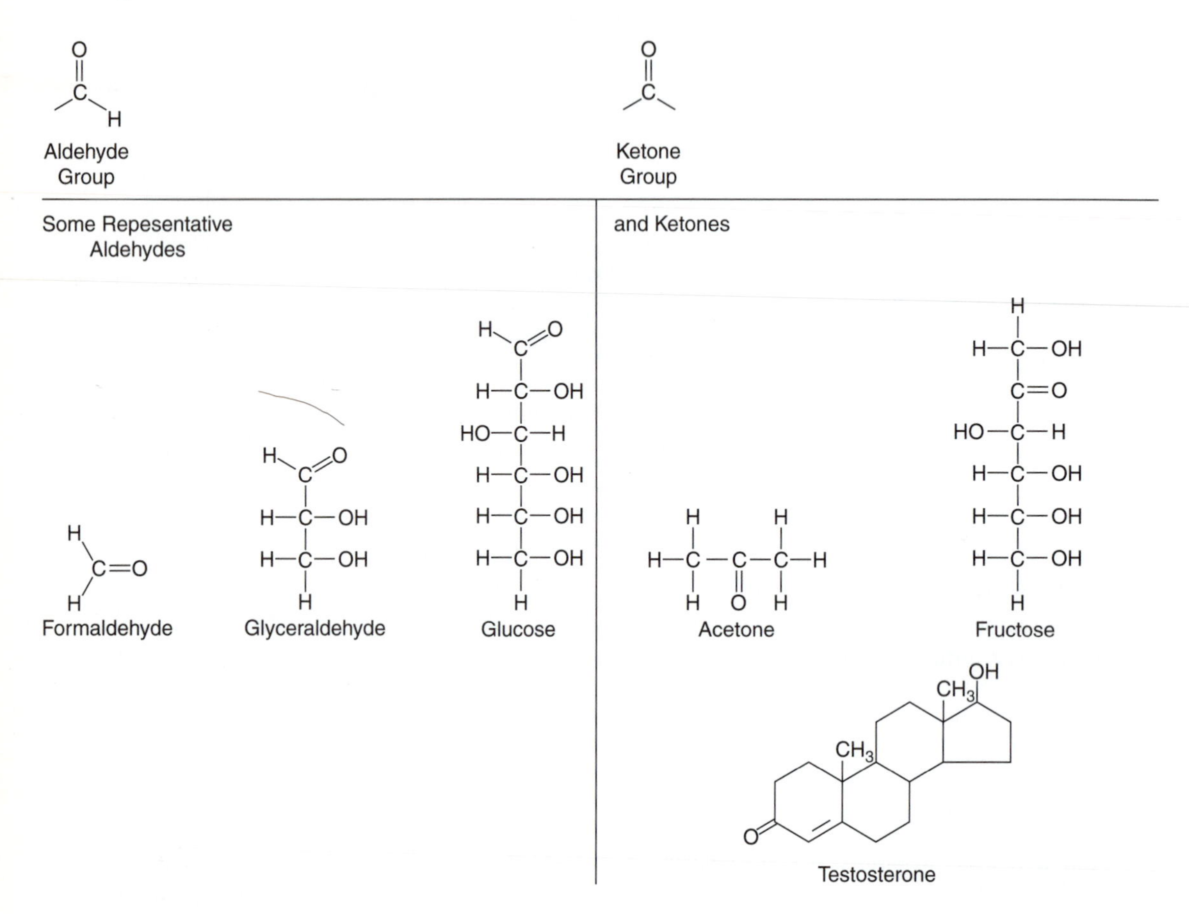

Questions

57. Circle the carbonyl group in all the aldehydes in the diagram above. Where is the carbonyl group located?

58. In aldehydes, the carbon that participates in forming the carbonyl group is also always bonded to ____________. Carbon can form ______ (how many) bonds. After it bonds to oxygen in a carbonyl group and bonds to ______, the end carbon in an aldehyde can form ______ more bond(s).

59. Circle the carbonyl group in all the ketones in the diagram above. Where is the carbonyl group located?

60. Aldehydes and ketones are similar because they both contain a ____________ group.

61. Explain how aldehyde and ketone groups differ from each other.

62. Glucose and fructose are both sugars. The chemical formula for glucose is ____________, and the chemical formula for fructose is ____________.

63. Both glucose and fructose contain a carbonyl. Glucose is classed as ____________, whereas fructose is classed as ____________. In addition to the carbonyl group, both sugars contain ____________ (how many) alcohol groups.

64. Compounds with the same chemical formula and different structural formulas are called ____________. Are glucose and fructose isomers? Explain your answer.

65. The carbonyl group ______ (is or is not) polar. Predict whether small molecules containing only a carbonyl group are likely to be soluble in water, and explain your reasons.

66. Testosterone is the principle male sex hormone. It contains four attached, or fused, rings and is therefore classified as a ____________. Where is the carbonyl group located in the testosterone molecule? ____________.

67. Explain why testosterone is classified as a ketone and not an aldehyde.

68. Circle all the functional groups that are present in testosterone, and name them. Most of the testosterone molecule is composed of ____________. There are ______ (how many) polar groups in testosterone, and the rest of the molecule is ____________. Testosterone ______ very soluble in water.

Carboxylic Acids

The carboxyl group contains the atoms –COOH, always bonded in the pattern shown below. Organic compounds containing a carboxyl group are classed as carboxylic acids (organic acids). Examples of carboxylic acids are shown in the diagram below. The two oxygen atoms in the carboxyl group attract the electron cloud quite strongly, making the bond to the hydrogen atom partially ionic. That hydrogen therefore has some tendency to dissociate as an ion, and this is responsible for the acidic properties of these compounds.

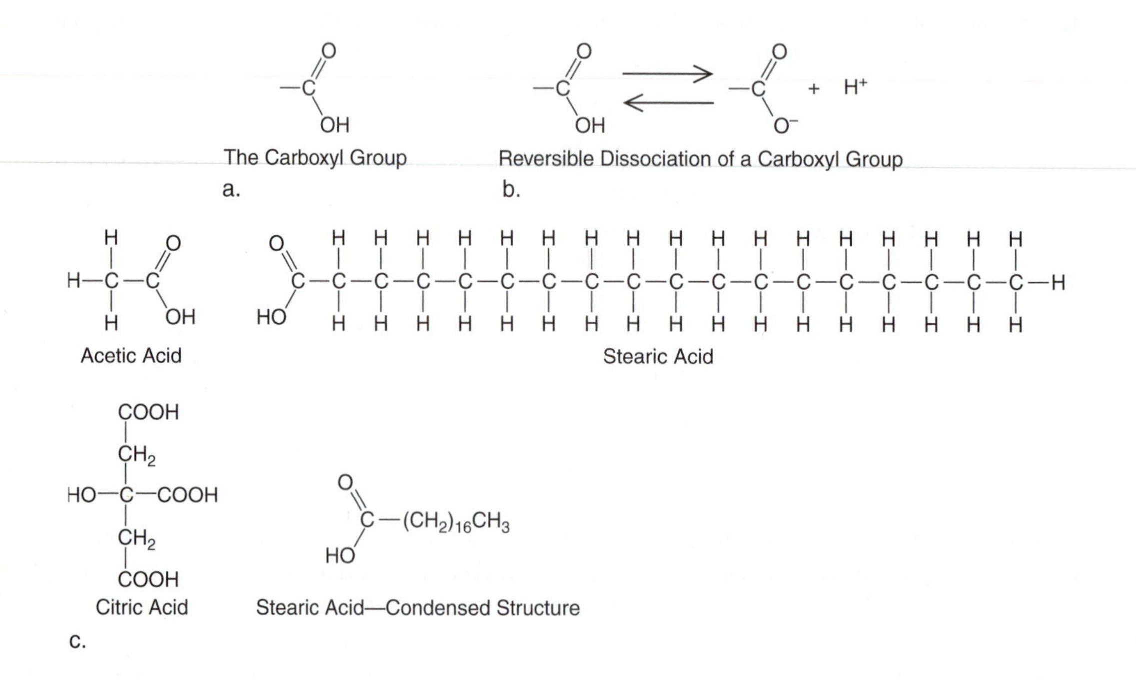

Questions

69. Oxygen is a very electronegative element. This means that the electron cloud will shift ____________ (toward or away from) oxygen and ____________ from carbon and hydrogen in the carboxyl group.

70. Covalent bonds that involve unequal sharing of electrons are called ____________. What is an ionic bond?

71. A covalent bond that is very polar is intermediate in character between __________________ (what other types of bonds). A bond of this type will have some tendency to dissociate and release an ion.

72. The bond linking hydrogen to oxygen in the carboxyl group is strongly polar and sometimes dissociates, releasing a _________________.

73. Acids are substances that release ___________ ions in solution. The –COOH group is ___________, because it can release hydrogen ions in solution.

74. Acetic acid is the acid in vinegar. Would you predict that acetic acid would be soluble in water? Explain the reasons for your prediction.

75. A solution of citric acid would be ___________ (acidic, basic, or neutral) due to the release of ___________ ions in solution. Citric acid is produced during the aerobic breakdown of glucose in most cells. Citric acid is present in larger amounts in citrus fruits such as grapefruit and lemons. The pH of lemon juice is ___________, and the juice tastes ___________.

76. The double arrow in the reaction for the dissociation of acetic acid means that the reaction proceeds in ___________ directions. At any time, part of the acetic acid is dissociated and much of it is not. Strong acids are completely dissociated into ions. Acetic acid ___________ a strong acid.

77.** A weak acid and the salt of the same weak acid (itself a weak base) can be mixed. Such mixtures tend to stabilize pH, and are called ___________. These mixtures tend to stabilize pH, because they can react with and neutralize both _________________.

78. Acetic acid and citric acid are more soluble in water than stearic acid. Explain why this is so.

79. Stearic acid is a member of a group of compounds called fatty acids. Most of the fatty acid molecule is ___________, and at one end there is a ___________. What does the subscript 16 mean in the condensed structure of stearic acid?

80. Carboxyl groups that have dissociated and released a hydrogen ion have a net charge of ___________. Structures that are charged due to ionization tend to be ___________ in water.

Amines

The amino group contains the atoms $-NH_2$, bonded in the pattern shown below. Organic compounds that contain this group are called amines, and some examples are shown in the diagram below. The nitrogen atom in the amine group can function as a base by accepting a hydrogen ion.

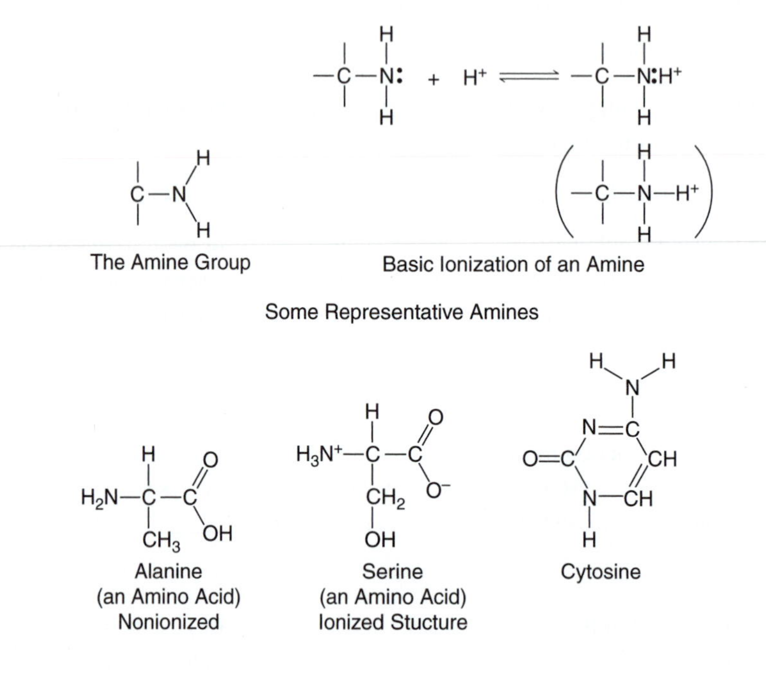

Questions

81. Nitrogen has ______ valence electrons. Nitrogen tends to form ___________ covalent bonds to create an octet in its valence shell.

82. Diagram the structure of ammonia. Show the shape and position of all valence electrons.

83. How many of the valence electrons of ammonia are involved in covalent bonding? ______ How many of the valence electrons of ammonia are not involved in covalent bonding? ______

84. Diagram the reaction of an ammonia molecule with a hydrogen ion to form an ammonium ion.

85. An amine group in an organic compound can function as an organic base, because it can accept a ______ ion on its unbonded pair of valence electrons.

86. What does the term "nitrogenous base" mean?

87. Alanine and serine are amino acids. What two functional groups are present in both these amino acids?

In addition to these two groups, serine also has a ____________ group and is therefore an alcohol.

88. In the cell, most amino acids carry a positive charge on the ____________ group and a ____________ charge on the carboxyl group.

89. Predict whether amino acids will be soluble in water, and explain the reasons for your answer.

90. Cytosine is one of the components that is present in DNA and RNA, which are large molecules that encode and express genetic information. Cytosine is a structure that contains a ring which is composed of carbon and ____________. The two functional groups attached to the cytosine ring are ____________ and ____________.

91. When an organic compound contains a ring composed of both nitrogen and carbon, the position of the nitrogen atoms is always shown. Are the positions of carbons always shown? Explain your answer.

Phosphates

The phosphate group is a derivative of phosphoric acid, H_3PO_4. One or more of the hydrogens in the phosphoric acid dissociates as a hydrogen ion at the pH values that are typical in living cells. Phosphate groups are therefore acidic. Examples of phosphates are shown in the diagram below.

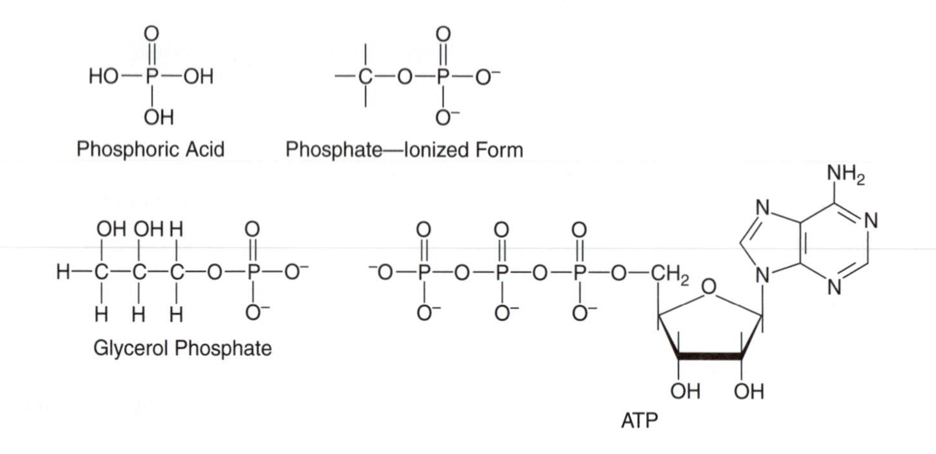

Questions

92. Phosphoric acid has the formula ____________. Phosphoric acid is acidic because it can release ____________ into solution.

93. After all three hydrogen ions dissociate, the remaining group is called ____________ and has a net charge of ____________.

94. One phosphate group can combine with ______ Na^+ ions to form a salt with the formula ____________, called ____________.

95. Covalent bonding of a phosphate group to carbon forms an organophosphate compound. Organic compounds that contain a phosphate group are likely to be ____________, because the hydrogen atoms within the phosphate are usually dissociated as hydrogen ions.

96. Glycerol phosphate is an organic compound that contains ______ (how many) phosphate group(s). Adenosine triphosphate (ATP) contains ______ phosphate group(s) and is ionized at physiological pH values. Negative charges ______ each other. Several negative charges are close together in the ________________ region of ATP.

Organic Chemistry: An Overview

Organic compounds exist in enormous variety, and you have now seen numerous examples. Although organic compounds are often complex, and there are millions of different compounds, you have already learned to recognize many of the unique organic "personalities," or functional groups, and can predict their behavior.

Organic chemists represent structures in ways that emphasize the identity of participating atoms and the location of bonds—very important when we are trying to determine what functional groups we have and the chemical character we expect. But what do these molecules really look like? When atoms bond to form molecules, their electron clouds blend to form new shapes. Space-filling models that represent the positions of the electron clouds of molecules are the most accurate way of representing molecular shapes. Some examples are shown in the figure below. Organic compounds interact with each other and with water through their shape and through attractions and repulsions of regions that are partially or fully charged. Notice the dissolved sugar with a personal escort of water molecules to which it is hydrogen bonded.

Only six elements are included within the organic functional groups you have learned to recognize, namely carbon, hydrogen, oxygen, nitrogen, phosphorus, and sulfur. Yet these six elements are sufficient to generate the complexity and variety we have begun to investigate. Complexity is generated by using this limited number of building blocks to assemble a great variety of structures. There are only a few kinds of building blocks, but the different ways they can be put together is essentially infinite. Carbon, with its valence of four and midrange electronegativity, plays a very special role. It functions as a versatile atomic link, the hook that can be used to combine many different pieces into an organic compound of unique shape and character. It is doubtful that life could have evolved without carbon. No other element produces the chemical variety necessary to provide the physical basis for the complicated phenomena of life.

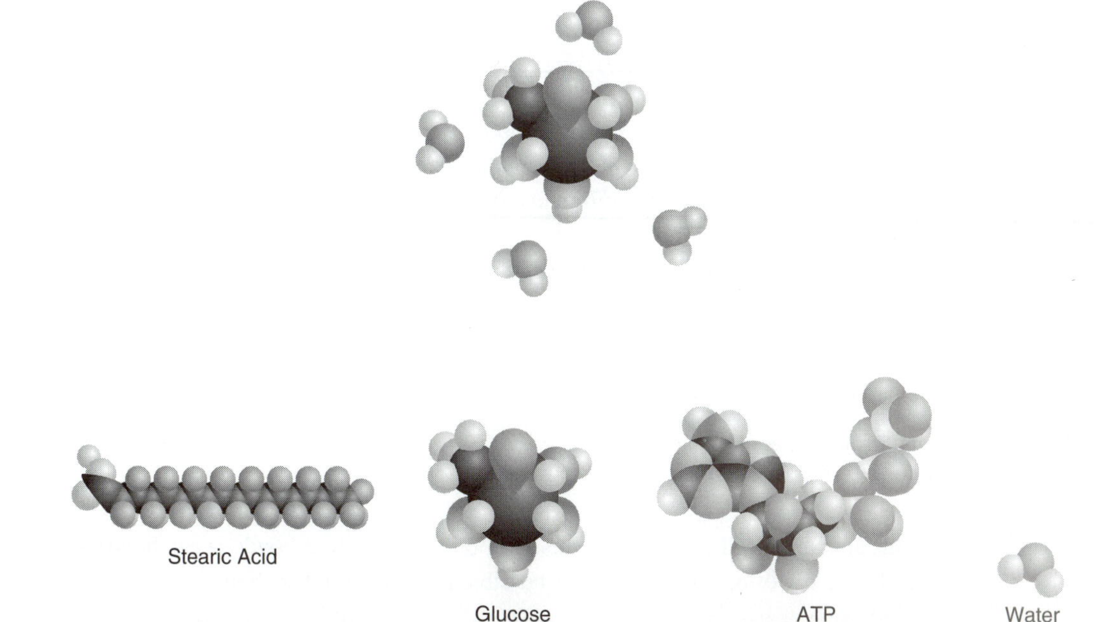

Answers

1. carbon.

2. less; water; organic compounds.

3. carbon.

4. four; share; four.

5. single covalent; two pairs (or four); triple; nonpolar (equal) covalent.

6. one; single covalent bond; one.

7.

Element	Electronegativity	Bond Formed with Carbon
C	2.5	Covalent—nonpolar
H	2.1	Polar covalent
O	3.5	Polar covalent
N	3.0	Polar covalent
S	2.6	Covalent—nonpolar

Explanations: Bonds between carbon atoms are nonpolar or equal, because all carbon atoms have the same electronegativity and will share electrons equally. Bonds between carbon and sulfur are considered nonpolar as well, because the difference in electronegativities is so small that no significant movement of the electron cloud is produced. Bonds between carbon and hydrogen are slightly polar, as are bonds between nitrogen and carbon. Bonds between oxygen and carbon are more strongly polar, because the difference in electronegativities of the two atoms is larger and there is a noticeable shift of the electron cloud toward oxygen.

8. organic functional groups.

9. Organic compounds are varied and complex, because carbon can form four bonds, it can bond to many other elements, and a variety of bonding patterns is possible in the carbon backbone.

10. equally; nonpolar covalent bonds; polar covalent bonds.

11. polar covalent.

12. four; intermediate; many.

13. ions; covalent bonds; do not; do not.

    Explanation: When compounds that contain ionic bonds dissolve in water, each ion moves into solution separately and is surrounded by a layer of water molecules. Because the ions are charged and can move independently in the solution, an electrical current can be conducted, and the solution is an electrolyte. Molecules that contain atoms linked by covalent bonds usually will dissolve in water if they are polar. Covalent bonds do not dissociate in water, and the compound dissolves as a unit, does not release independent charged particles, and is a nonelectrolyte.

14. carbon and hydrogen; CH_4; C_2H_6; hydrocarbons; do not; nonpolar.

15. hydrophobic; hydrophobic; No. Organic functional groups such as hydrocarbons have regular and predictable properties. Other hydrocarbons will therefore probably have properties similar to methane and ethane. They will be hydrophobic and fail to dissolve in water because they are nonpolar.

16. Butane, pentane, and pentene do not branch. Isobutane does branch.

17. The carbon chain in cyclohexane and benzene forms a ring.

18. Pentene has a double bond between two of the carbons in the backbone. All the carbons in pentane are linked by single bonds.

19. Carbon can form four bonds. The four bonds can be any combination of single, double, and triple bonds, as long as the total is four. Carbon atoms bond readily to each other in straight chains, branched chains, and closed rings. This variation in the carbon backbone of organic molecules results from the fact that carbon can form four bonds.

20. C_4H_{10}; C_4H_{10}; are not; are.

21. two. Carbon can form a total of four covalent bonds. An internal carbon uses two bonds to link to the carbons on each side. It can form only two more covalent bonds to complete its valence shell and thus bonds to just two hydrogens. A carbon on the end of the molecule bonds only to one other carbon and can therefore bond to three hydrogens.

22. nonpolar; hydrophobic.

23. releases; fuels.

24. energy (or heat); hydrocarbons.

25. does not; nonpolar; hydrophobic; on top of.

26. The carbon atoms in the cyclohexane molecule are linked by single covalent bonds. Each carbon is bonded to two other carbon atoms and to two hydrogen atoms to reach its valency of four. The benzene ring contains three double bonds and three single bonds. Each carbon forms three bonds to other carbon atoms and therefore bonds to only one hydrogen.

27. Both contain a ring formed of six carbon atoms. The compounds differ because cyclohexane has no double bonds and therefore contains more hydrogen (it is "saturated" with hydrogen). Benzene does contain some double bonds and therefore contains less hydrogen (it is "unsaturated" with hydrogen).

***Cyclohexane is not a planar or flat molecule. When a carbon atom forms four single bonds, the geometry of that carbon is tetrahedral, with carbon at the center of the tetrahedron. The cyclohexane molecule is flexible and can have the shapes shown below. These shapes allow each carbon in the ring to approximate tetrahedral geometry. When a carbon atom forms a double or triple bond, the geometry is changed and the shape at that carbon is not tetrahedral. This leads to different molecular shapes.*

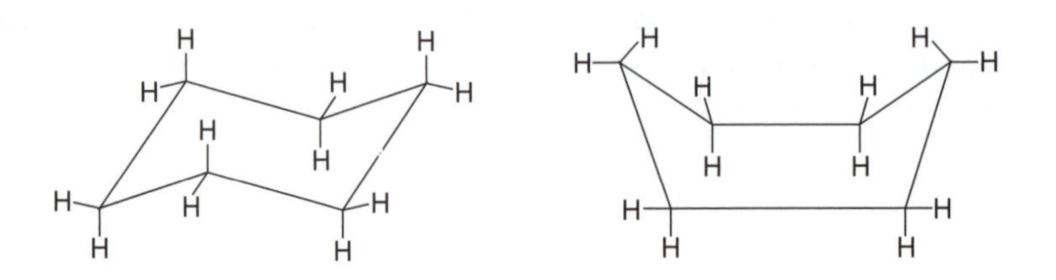

28. is.

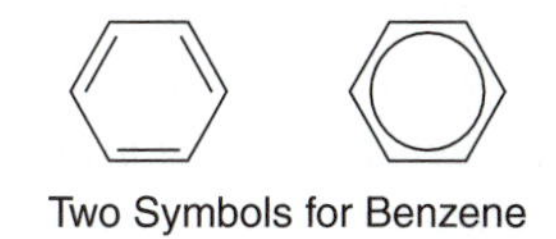

Two Symbols for Benzene

29. are; ultraviolet light.

30. are; are not.

31. are not; are not.

32. are not; are not.

33. Carbon always has a valence of four. Determine how many bonds are present linking carbon atoms to each other. The carbon will then bond to the number of hydrogens needed to reach the total of four bonds. For example, each carbon in the benzene ring participates in a double and a single bond to other carbon atoms. It therefore can bond to only one hydrogen to make a total of four bonds.

34. Lines and geometric shapes are used to represent some organic molecules. If atoms are not shown, a carbon is understood at each angle in the geometric shape or where lines join or form an angle in a line drawing.

35. do not; hydrophobic; form a layer on top of the water; high.

36. *b.* lipids such as fats and oils.

37. polar; more.

38. hydrophobic; polar; hydrophilic.

39.* *Hydrogen bonds are weak attractions between different molecules. Water molecules can hydrogen bond to each other and tend to stick together in clusters (it's an associated liquid). A solute must make a space for itself in this "sticky" water. That happens more readily if the solute can form hydrogen bonds with water. A solute that cannot form hydrogen bonds must make a kind of cavity for itself in the sticky water by breaking some hydrogen bonds between water molecules. If it does not replace these broken bonds with equivalent interactions, it will tend to be less soluble than a substance that can form hydrogen bonds or other ionic interactions.*

40. weak; do not; Hydrogen bonds can form when hydrogen is present in a strongly polar covalent linkage, thus developing a partial positive charge. Bonds linking hydrogen to oxygen or nitrogen are sufficiently polar. There must also be an element with a partial negative charge, such as oxygen or nitrogen. The partial charges of opposite sign are attracted to each other, forming a hydrogen bond.

41. can; increase.

42. Ethanol is more soluble in water, because it contains a polar group (hydroxyl) that can also form hydrogen bonds with water. Ethane is nonpolar and is not soluble.

43. covalent (and polar); does not. Alcohols are nonelectrolytes, because they do not dissociate into ions when they dissolve in water. An electrolyte must have mobile charged particles in solution. An alcohol is not charged.

44. -ol; the hydroxyl or alcohol.

45. five; Glucose dissolves in water because it has five alcohol groups. Alcohol groups are polar and can form hydrogen bonds with water, promoting solubility. Large amounts of sugar will dissolve in water.

46. hydroxyl or alcohol.

47. hydrocarbon; hydrogen bonds; are not; hydrocarbon.

48. Cholesterol is not soluble in water. Although it contains one polar group, the great majority of the molecule is hydrocarbon, which dominates the properties of the molecule as a whole.

49. hydrocarbon; hydrogen.

50. hydrocarbon; dissolve in water.

51. The hydrocarbon chain that is attached to the five-member ring is represented with a line structure in the abbreviated formula. A carbon with the appropriate number of hydrogens is understood to be present at each angle.

52. two; octet; two; two.

53. Many sulfur-containing compounds have a distinct odor. Humans (and many other animals) have chemical receptors that can detect these compounds.

54. less; do. Thiols do not form hydrogen bonds, because the covalent bond between sulfur and hydrogen is not polar enough. Since the hydrogen does not develop a significant partial charge, hydrogen bonds cannot form.

55. cysteine; thiol.

56. strong; do not.

57. The carbonyl group is located on the end of the molecule in aldehydes.

58. hydrogen; four; hydrogen; one.

59. The carbonyl group in a ketone is located in the internal part of the molecule (anywhere except on the end).

60. carbonyl.

61. Aldehydes and ketones differ, because the carbonyl group occupies a different position. The carbonyl group is on the end of the molecule in aldehydes, and is located within the carbon chain in ketones.

62. $C_6H_{12}O_6$; $C_6H_{12}O_6$.

63. an aldehyde; a ketone; five.

64. isomers. Yes, glucose and fructose are isomers. They have the same chemical formula ($C_6H_{12}O_6$) but differ because the atoms are connected in a different pattern (they have different structural formulas).

65. is. Aldehydes and ketones that do not contain large hydrocarbon regions do tend to dissolve in water, because of the polar carbonyl group.

66. steroid. It is attached to one of the six-member rings in the steroid ring system.

67. The carbon bearing the carbonyl group is also bonded to two carbons, one on either side. This qualifies as a ketone. An aldehyde must have a carbonyl group attached to a carbon that also is bonded to hydrogen. The carbonyl group in testosterone does not meet this definition.

68. hydrocarbon; two; hydrocarbon; is not. Testosterone contains a ketone group (on a six member ring) and an alcohol (on the five-member ring)—these groups should be circled.

69. toward; away.

70. polar. An ionic bond is the attractive force between ions of opposite sign. Ions form when electrons are transferred from an element that functions as an electron donor to one that can accept electrons.

71. covalent and ionic.

72. hydrogen ion.

73. hydrogen; acidic.

74. Yes. Acetic acid contains the carboxyl group, which is strongly polar. Some of the carboxyl groups ionize, releasing H^+. Polar substances and ions are usually soluble in water.

75. acidic; hydrogen; below 7; sour.

76. both; is not.

77.** buffers; strong acids and strong bases.

78. *Stearic acid contains a long hydrocarbon chain in addition to one carboxyl group. Solubility depends on the proportions of the molecule that are hydrophilic and hydrophobic. Stearic acid is not very soluble in water, because the long hydrocarbon chain in stearic acid is hydrophobic and only the carboxyl group is hydrophilic. Acetic acid and citric acid do not contain large hydrocarbon regions that would decrease solubility in water.*

79. hydrocarbon; carboxyl group. The subscript 16 by the (CH_2) group means that sixteen copies of the CH_2 group are present in stearic acid, linked together by covalent bonds.

80. –1; soluble.

81. five; three.

82.

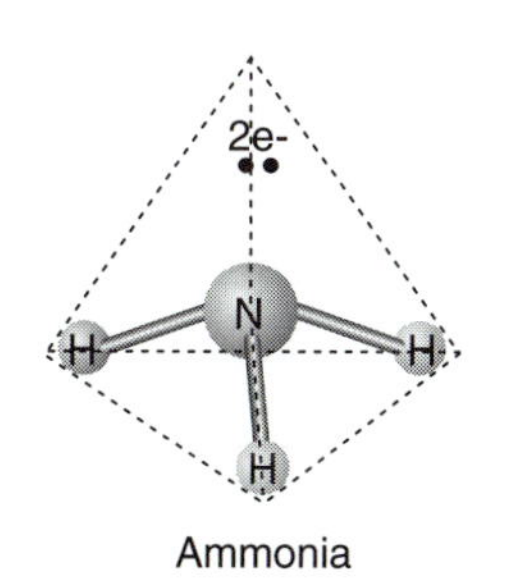

Ammonia

83. six (three pairs); two (one pair).

84.

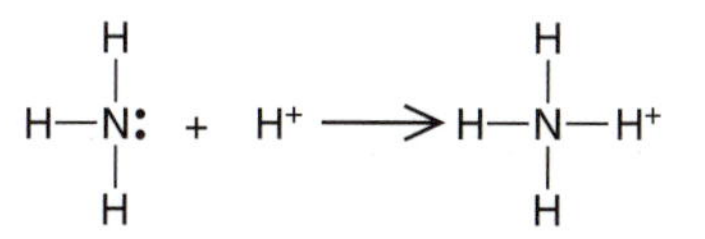

85. hydrogen.

86. A nitrogenous base is an organic compound that contains nitrogen. Nitrogen in the organic compound can react with hydrogen ions and thus functions as a base.

87. Both amino acids contain an amine group and a carboxyl group; hydroxyl.

88. amino; negative.

89. Amino acids are soluble in water, because they have groups that are either very polar or ionized. This promotes solubility in water.

90. nitrogen; an amine and a ketone.

91. No. Some abbreviated or condensed structures represent the position of carbon atoms but do not list them directly.

92. H_3PO_4; hydrogen ions.

93. phosphate; –3.

94. three; Na_3PO_4; sodium phosphate.

95. ionized.

96. one; three; repel; triphosphate tail of ATP.

Note: ATP is a molecule that stores energy. Living cells use ATP as an energy transfer molecule. ATP delivers energy to processes such as muscle contraction that require it.

Chapter Test: Organic Chemistry

The questions in this chapter test evaluate your mastery of all the objectives for this unit. Although material from earlier chapters is not tested directly, recall of concepts from earlier chapters is necessary in some cases. Take this test without looking up material in this or previous chapters. You may use the periodic table. After you have completed the test, you can check your work by using the answer key located at the end of the test.

1. Organic compounds are always based on the element ____________.

2. Organic chemistry is complex and varied, because carbon has such a versatile bonding pattern. Carbon exhibits this variety, because it can form ______ bonds, has an ____________ electronegativity, and can therefore bond to ______ other elements.

3. Organic compounds that contain only hydrogen and carbon are called ____________, have very ______ solubility in water, and are therefore ____________.

4. The –COOH functional group is called a(n) ____________ group. Compounds that contain this group tend to dissolve in water and are ____________ (acidic, basic, or neutral), because the group releases ____________ in solution.

5. The –OH functional group is called a(n) ____________ group. Compounds that contain this group tend to dissolve in water and are ____________ (acidic, basic, or neutral), because the group releases ____________ in solution.

6. Which organic functional group exhibits basic behavior in aqueous solution? ____________ Explain how this group functions as a base.

7. Amino acids are the building blocks of proteins. All amino acids contain a(n) ____________ group and a(n) ____________ group. Amino acids ____________ electrolytes, because they ____________.

8. An organic molecule that contains a carbonyl group on the end of the molecule is a(n) ___________. If the carbonyl group is located on an internal carbon, rather than the end, the organic molecule is a ___________.

Questions 9–16 are based on the structures in the diagram below.

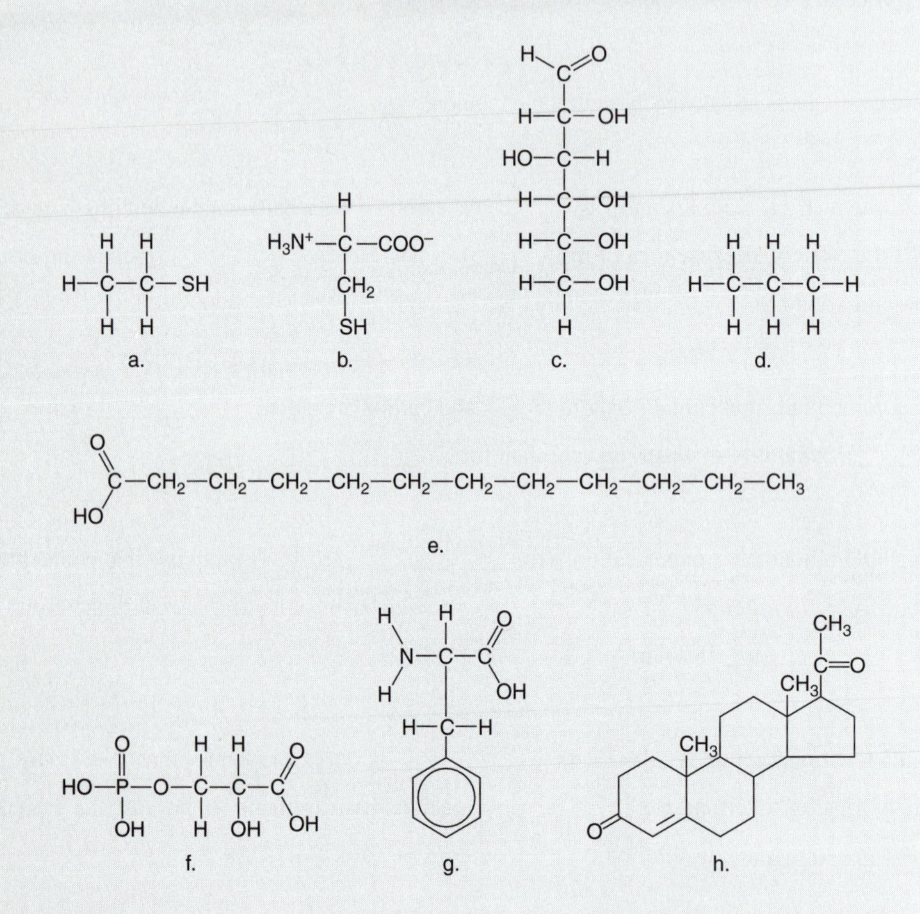

9. An alcohol contains the ______ group. Which of the compounds are alcohols? ______

10. Ketone groups are present in ___________. Circle the ketone group(s).

11. Which of the compounds are amines? ___________. Explain why the amine group is written in two different ways in the compounds that contain it.

12. Three of the compounds are not very soluble in water. Predict which three would be least soluble and explain why.

13. A benzene ring is present as part of the structure of compound ____________.

14. List all the functional groups that are present in compound *f*. __________________ Would compound *f* dissolve in water? ____________. Predict whether it will change the pH of the solution, and explain your reasoning.

15. Compounds ____________ are thiols due to the presence of the ____________ group.

16. Which compound(s) are amino acids? ____________ Are there any other functional groups (in addition to the carboxyl and the amine groups) present in these amino acid(s), and if so, which ones? ____________

17. List the organic functional groups that are polar but nonelectrolytes when dissolved in water.

18. List the organic functional group(s) that are hydrophobic, and explain why they are hydrophobic.

19. List the organic functional groups that are hydrophilic and are electrolytes when dissolved in water. __________________ Explain why these groups are electrolytes, whereas the groups listed in question 17 are not.

20. Benzene has the formula C_6H_6. It is often represented by abbreviated symbols. What are these symbols and what do they mean?

Answers

1. carbon.

2. four; intermediate; many.

3. hydrocarbons; low; hydrophobic.

4. carboxyl; acidic; hydrogen ions (H^+).

5. hydroxyl; neutral; no ions.

6. The amine group.

   Amine groups function as bases, because the nitrogen can bond with a hydrogen ion (H^+). The hydrogen ion shares the pair of nonbonding valence electrons that is present in the nitrogen. Any substance that can accept (or react) with H^+ is a base. Organic compounds that contain amine groups are often referred to as "nitrogenous bases."

7. amine; carboxyl; are; release ions in aqueous solution.

8. aldehyde; ketone.

9. hydroxyl; *c* and *f* are alcohols, due to the presence of –OH groups bonded to carbon.

10. Compound *h* contains two ketone groups as shown below.

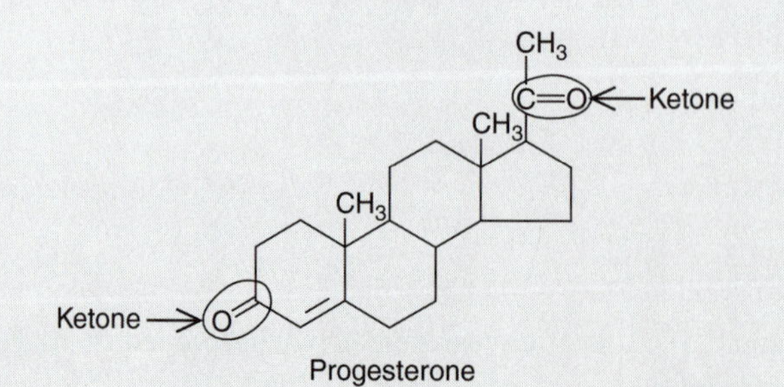

11. Compounds *b* (cysteine) and *g* (phenylalanine) contain the amine group and are classed as amines.

    In cysteine, the amine group is written as H_3N^+–. This notation shows that the nitrogen has accepted a hydrogen ion and now carries a charge of +1. In contrast, the amine group in phenylalanine is shown in the nonionized form –NH_2 (the amine group has not yet acted as a base). In cells, most amine groups are ionized.

12. Compounds *d, e,* and *h* would be hydrophobic.

    Compound d is composed entirely of hydrocarbon. Hydrocarbons are nonpolar and therefore have little tendency to dissolve in a polar compound such as water. Although other groups are present, compounds e and h are formed predominantly of hydrocarbon, and are therefore hydrophobic as well.

13. *g*

14. Compound *f* contains a phosphate group, a hydroxyl (alcohol), and a carboxyl group. Yes. When compound *f* (phosphoglyceric acid) dissolves in water, it lowers the pH. Both the carboxyl (–COOH) and the phosphate (–H_2PO_4) can go through acid ionization by releasing hydrogen ions in solution. Release of H^+ will raise the acid concentration and lower the pH.

15. *a* and *b*; sulfhydryl (–SH).

16. *b* and *g*. Yes, *b* contains a thiol group, and *g* contains a benzene ring.

17. Alcohols, carbonyls (aldehydes and ketone), and thiols are polar but do not ionize in solution. They therefore are nonelectrolytes.

18. Hydrocarbons are hydrophobic, because the molecules as a whole are nonpolar. They cannot form hydrogen bonds with water and do not tend to dissolve.

19. Amines, phosphates, and carboxyl groups are hydrophilic and are electrolytes. Certain bonds within the functional groups are so polar that they have some ionic character and tend to partially dissociate in water, freeing ions. Any substance that releases ions in solution is an electrolyte.

20.

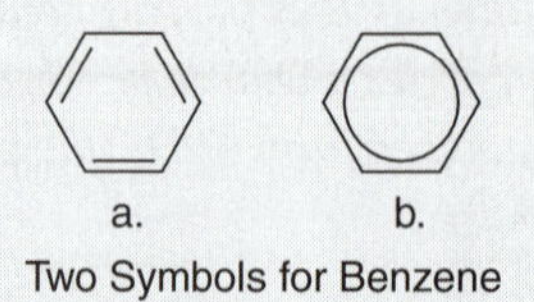

Two Symbols for Benzene

The hexagon represents a six-carbon ring. The alternating double and single bonds in benzene are represented with double and single lines within the ring in a. The six hydrogens that are necessary to complete the valence requirements of carbon are understood to be present, and are not shown. The circle inside the hexagon in b represents the electrons as delocalized, a cloud of electrons that belongs to the entire benzene ring, rather than as alternating double and single bonds with definite locations. Both formulas appear to be flat, as is the benzene molecule.

Chapter 6

Biomolecules

Objectives

1. Define the following terms and apply the definitions correctly: dehydration synthesis, digestion, heteropolymer, homopolymer, hydrolysis, macromolecule, monomer, and polymer.
2. List the four classes of biomolecules that are most prevalent and most important in living organisms. Know the monomers and polymers in each class.
3. List the functions performed by proteins, carbohydrates, nucleic acids, and lipids.
4. Explain how peptide bonds are formed. Describe the primary, secondary, tertiary, and quaternary levels of structure of a protein.
5. Explain the differences between protein hydrolysis and denaturation.
6. Define complementary fit, ligand, and conformation change. Explain how these concepts are related to protein specificity and normal protein function.
7. Explain how glycosidic linkages form.
8. Define the terms used to describe carbohydrates: triose, pentose, hexose, monosaccharide, disaccharide, and polysaccharide.
9. List the common polysaccharides, state their function, and explain how they differ structurally from each other.
10. List the common lipids, state their function, and explain how they differ structurally from each other.
11. Explain how ester linkages form.
12. Explain how saturated fatty acids differ from unsaturated fatty acids.
13. Define these terms as they apply to lipids: dehydrogenation, hydrogenation, phospholipid bilayer, and triglyceride.
14. List the components that are present in a nucleotide.
15. List and explain the ways that RNA and DNA differ.
16. Define these terms as they apply to nucleic acids: base pairs, double helix, polynucleotide, sequence variation, and sugar phosphate backbone.
17. Explain how DNA codes information.
18. Explain the importance of ATP to cells and organisms. Explain how ATP functions.

Organic Molecules in Living Organisms

Water is the most prevalent compound in cells and organisms. The dry weight of an organism is the matter that remains after water is removed. Analysis of the dry weight derived from many living organisms shows that it is composed mostly of organic compounds. The organic compounds can be classed into four major categories: proteins, carbohydrates, lipids, and nucleic acids.

Questions

1. When leaves from a plant are dried, the weight of the leaf tissue ____________ a great deal, because ______ is the most plentiful compound in leaves and most other samples of living tissue.

2. A dehydrated or dried sample of blood weighs much less than the original fluid sample, because ____________ is the most abundant compound in blood.

3. Most of the dry weight of the leaf tissue is composed of ____________ compounds, all of which are based on the element ____________.

4. Analysis of dried blood residue, or of the dry material from leaves, shows that a small amount of matter—such as sodium ions, potassium ions, chloride ions, and inorganic phosphates—is present. Matter of this type is classed as ____________ and would be a small proportion of the dry weight of the tissue. Most of the dry weight of the tissue is composed of ____________.

5. Cellulose is a carbohydrate that is plentiful in plant tissue. Is cellulose organic or inorganic? Explain your answer.

6. Dehydrated animal tissues contain mostly protein, some fat (a lipid), and small amounts of nucleic acid and carbohydrate. All these compounds are ____________ and are based on the element ____________. Dehydrated animal tissues also contain some ionic materials, such as Na^+, K^+, and Cl^-, that are ____________ and are not based on carbon.

Many of the organic molecules that are present in cells are extremely large, with molecular weights in the thousands or millions. These large molecules, called macromolecules, have complex properties, yet their structure is quite repetitive. Macromolecules are assembled by linking building-block molecules into long chains, much as a necklace is assembled by stringing beads. Molecules that are used as building blocks are called monomers (mono means one). A long chain of linked monomers is called a polymer (poly means many).

$$\text{Monomers} \rightleftarrows \text{Polymers + many water molecules}$$

Questions

7. A ____________ is an extremely large molecule that can be produced by linking many building-block molecules, called ____________, together. The chemical reaction that links the small molecules also produces ____________.

8. The prefix "oligo" means a few. If several thousand monomers are linked together, the products are a ____________ along with several thousand ____________. If 10 monomers are linked together, the product would be an ____________ and a few water molecules.

9. Anything that can be synthesized in a living cell can also be broken down or digested. Digestion of a polymer produces many ____________.

10. A protein is a polymer composed of linked amino acids. The amino acids are the ____________ that can be linked to form many water molecules and a ____________. Digestion of a protein would release ____________.

11. Starch is a macromolecule, or ____________, composed of many linked glucose units. Starch is a homopolymer, a term that means all the monomers within it are the same. Digestion of starch produces many ____________ molecules.

12. When protein is digested, 20 different kinds of monomers, called ____________, are produced. Protein is a ____________ (heteropolymer or homopolymer), because the twenty different amino acids have different structures.

13.* Predict whether starch molecules or proteins would be more variable in structure and function, and explain your answer.

The four classes of organic compounds (biomolecules) that are most important in living organisms are listed below. Protein, starch, cellulose, DNA, and RNA are all macromolecules.

Group	Monomers	Polymers
Proteins	Amino acids	Peptides and proteins
Carbohydrates	Sugars	Starch and cellulose
Nucleic acids	Nucleotides	DNA and RNA
Lipids	Fatty acids & glycerol	Fats & oils*

*One important group of lipids, the fats and oils, are composed of three fatty acids linked to one glycerol molecule. Although they are large molecules, they are not large enough to classify as polymers (macromolecules).

Questions

14. DNA and RNA are heteropolymers. When they are digested, they release ____________ type of monomer. The monomers that are linked to form DNA and RNA are called ____________.

15. The largest macromolecule in living cells is DNA. When DNA is digested, it releases large numbers of ____________, which ____________ (are or are not) all the same.

16. Genes are molecules that carry coded information that is essential to the cell. Genes are composed of DNA, a __________________ (hetero- or homopolymer) formed by linking four different kinds of __________________ together.

17. DNA, RNA, and proteins show more variation in their structures than ____________. DNA, RNA, and proteins are more variable because they are __________________, (heteropolymers or homopolymers) whereas ____________ is a homopolymer.

18. Structure and function are always closely related. Predict whether proteins or starches would have a greater variety of biological functions, and explain the reasons for your prediction.

19. A peptide (short for oligopeptide) releases a few ____________ when it is digested, whereas a protein (a polypeptide) would release many more ____________ after digestion.

20. Linking a glycerol molecule to three fatty acids produces a ____________. Though fairly large, this molecule is not a macromolecule or ____________.

Monomers are linked together to form dimers by a reaction called dehydration synthesis (also called condensation synthesis). The reverse reaction is called hydrolysis. A general example of these reactions is shown below. This type of reaction can be repeated many times to form polymers.

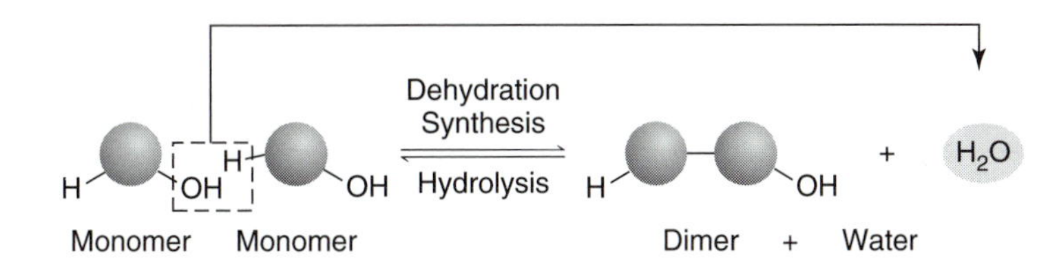

Questions

21. The reactants for a reaction are shown on the ______ side of a chemical equation, and the products are shown on the right side. The reaction that links monomers to form a dimer and a molecule of ____________ is called __________________.

22. Hydrolysis is a term that means to "break with water." ____________ is the chemical reaction that is important in digestion. A polymer and many water molecules can be converted to ____________ if the process of hydrolysis is complete.

23. Hydrolysis of proteins produces a mixture of ____________, and hydrolysis of DNA produces a mixture of ____________. Hydrolysis of starch produces only the sugar ____________, so starch is classified as a __________________ (heteropolymer or homopolymer).

24. Neurotransmitters are chemical messengers that are important in the communication between neurons (specialized cells in the nervous systems of most animals). Some neurotransmitters are oligopeptides. Synthesis of a neuropeptide would require several ____________, and the reaction that would link them together is called ____________.

25. Insulin is a hormone composed of protein. Hydrolysis of insulin would release ____________.

26. One of the monomers in a dehydration synthesis reaction contributes a(n) ____________ group, and the other monomer contributes a ____________. These components make up the water molecule, which is always produced by a dehydration synthesis reaction.

27.* Synthesis of a polymer requires that many monomers link together by ____________ reactions. Therefore, there must be at least two functional groups in each monomer: one that can contribute ____________ groups, and one that can contribute ____________.

Proteins

All cells contain proteins, the most versatile of the biomolecules. Proteins perform many essential tasks within cells, as the table below summarizes.

Function	Example and Description
Catalysis	Enzymes are fast, specific biological catalysts that greatly speed chemical reactions within cells.
Carriers	Molecules that are not very soluble in water are often transported by specific proteins. Hemoglobin binds and transports oxygen.
Communication	Receptors in cell membranes and some hormones are proteins.
Defense	Proteins of the immune system detect foreign molecules and may attack with specific proteins called antibodies.
Movement	The contractile proteins in muscle and other "protein motors" produce movement.
Structure	Various insoluble protein fibers strengthen the cell and make up most of the skin and much of some connective tissues such as cartilage, tendons, and ligaments.

Questions

28. Proteins are present in ______ cells and perform many essential functions.

29. Catalysts are compounds that ____________ chemical reactions. Biological catalysts are called ____________ and are composed mainly of ____________.

30. Compared to inorganic catalysts, enzymes are usually faster and ____________.

31. Many living organisms use oxygen very rapidly, yet oxygen has limited solubility in water. Some animals bind oxygen to a special protein called ____________, which carries oxygen throughout the body and delivers it to tissues. The protein ____________ makes it possible for cells to have more oxygen than would be available without its help.

32. Fats ______ (are or are not) soluble in water, because they are composed mostly of ____________. In animals, fats are transported by ____________.

33. Some proteins form insoluble, strong fibers. These proteins are used as ________________________.

34.* Structure and function are always related. Try to explain how one class of biomolecules, the proteins, could have enough structural variety to be suitable for all the different jobs they perform in cells.

Amino acids are monomers that can link together to form proteins. A generic amino acid and four specific examples are shown below.

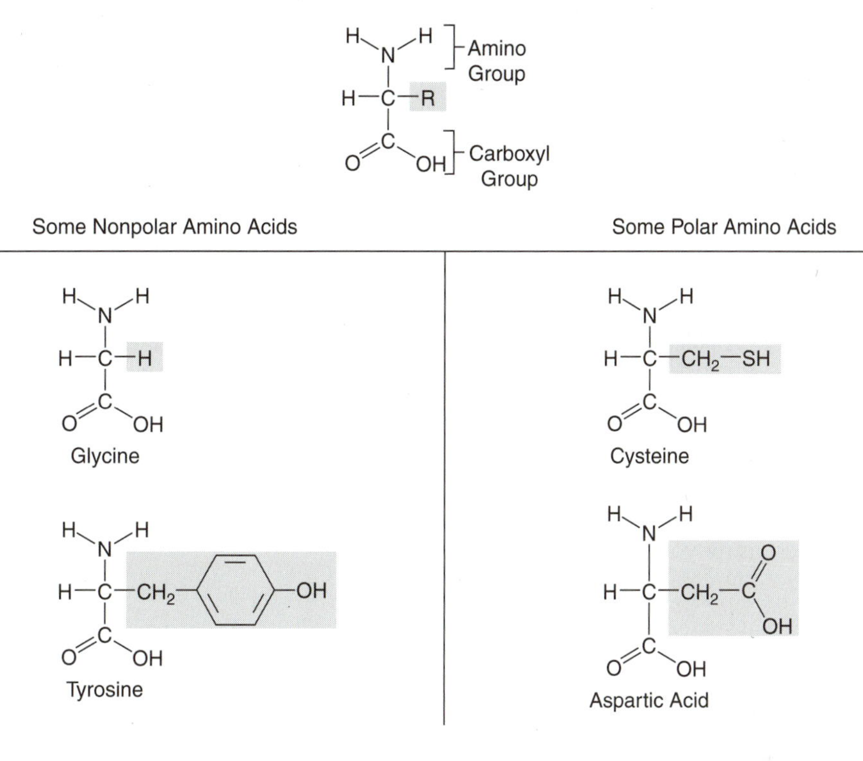

Questions

35. What two functional groups are always present in amino acids?

36. Which groups in amino acids can ionize at the pH within a typical cell? ____________________
Explain how these group(s) ionize.

37. A weak acid is _____________ (partially or completely) ionized and releases _____________ ions into an aqueous solution. A weak base can accept and react with _____________. A mixture of a weak acid and a weak base can help to stabilize pH and is called a _____________.

38. Amino acids (and proteins) can function as ____________, because they can both release and accept ______.

39. There are 20 unique amino acids that are found in naturally occurring proteins. How do the 20 amino acids differ from each other?

40. The part that differs in each of the 20 amino acids is called the variable group. It is often represented with the letter R. Variable groups may be composed of hydrocarbon, and the variable portion of the amino acid tends to be ____________. If the variable group has a structure that is polar or ionized, it will be ____________ and will have a tendency to promote solubility of the amino acid in water.

41.* Would amino acids be soluble in water? ____________ Would some be more soluble than others? Explain the reasons for your answer.

The dehydration synthesis reaction that links amino acids together to form peptides and proteins is shown below. The amino group and the carboxyl group both participate in this reaction.

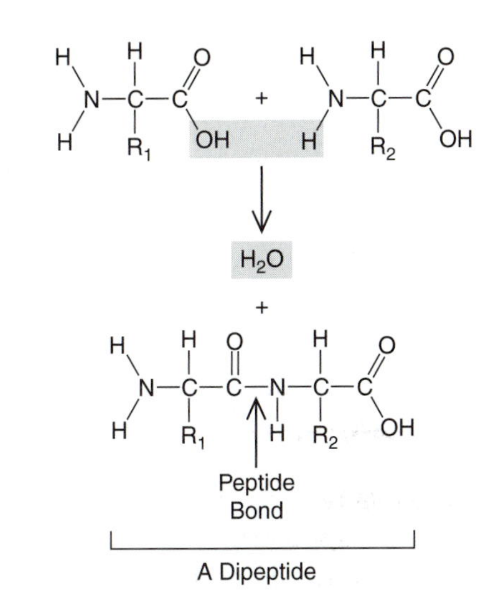

Questions

42. Which functional group in the amino acid contributes the hydroxyl group that ends up in the water molecule? ____________ Carbon always forms ______ (how many) covalent bonds. After the bond linking the hydroxyl group to carbon is broken, carbon is bonded to only ______ (how many) groups and is now reactive and unstable.

43. The ____________ group is the source of the hydrogen that ends up in the water molecule. Nitrogen usually forms ____________ (how many) covalent bonds to complete an octet in its valence shell. After the bond linking hydrogen to ____________ is broken, the nitrogen is bonded to ____________ groups and is reactive and unstable.

44. Two amino acids are linked together after the dehydration synthesis reaction is completed. The new bond that links them is called a ____________, and it joins a carbon atom in the first amino acid to a ______ atom in the second amino acid. The molecule that has formed by bonding the two amino acids together is called a ____________.

45. If another amino acid reacted by dehydration synthesis with the dipeptide shown on the previous page, the products would include ____________ and another ____________. The structure made up of three amino acids bonded together would contan ______ (how many) peptide bonds.

46. Ten amino acids linked together by dehydration synthesis would form a(n) ____________ and ____________ (how many) water molecules. There would be ____________ peptide bonds linking the amino acids together.

47. Peptide bonds are strong covalent bonds that link ____________ into peptides and proteins. Peptide bonds can be broken during the reaction called ____________. The products of hydrolysis are ____________.

48. The most important reaction in digestion is ____________. Large molecules such as peptides and proteins cannot be absorbed from the digestive tract, but smaller molecules, such as the ____________ released after hydrolysis, are absorbed. In the cell, both the reactions of dehydration synthesis and the reactions of hydrolysis occur with the help of enzymes, ____________ that speed the rate of chemical reaction.

The primary structure of a protein or peptide is the sequence of amino acids within it. The primary structure is specific and unique to each protein. Each primary structure is controlled by a gene, and there is one gene for each type of polypeptide. The gene specifies the length of the polypeptide and the sequence of amino acids within it.

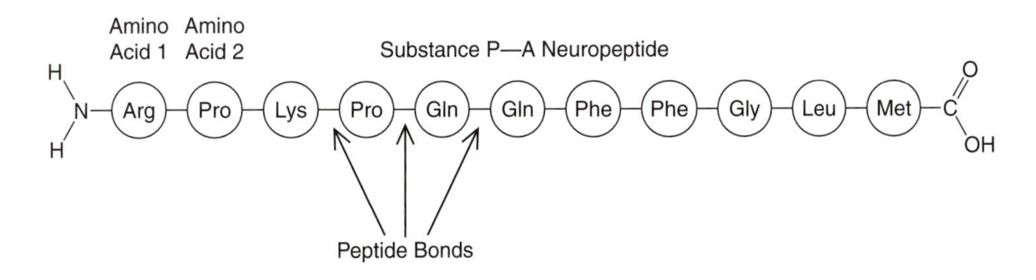

Questions

49. The primary structure shown above is that of Substance P, a neuropeptide that is involved in transmission of pain signals in the nervous system. Substance P contains ______ (how many) amino acids linked together by ______ peptide bonds. Substance P is called a(n) ____________. One end of the Substance P molecule ends with a(n) ______ group, and the other end has a carboxyl group.

50. Lysozyme is a protein that functions as an enzyme. Lysozyme contains 129 amino acids linked together by ______ peptide bonds. The sequence of amino acids in lysozyme is always the same and is controlled by ____________. The gene that provides the information for building lysozyme molecules controls the number of amino acids that are used, and the ________________.

51. Lysozyme and Substance P can be digested by a reaction called ____________. These reactions require water and a(n) ____________ to speed the chemical reaction. Peptide bonds are strong covalent bonds and are ____________ (stable or unstable). Enzymes that digest proteins are called proteases. Proteases speed the reaction of ____________.

52. The primary structure of a protein is ________________________. The primary structure of a protein is controlled by ____________.

53. The primary structure of collagen, a fibrous structural protein found in connective tissue, would be __________________ (the same, or different from) the primary structure of lysozyme. It would have a unique sequence of ____________, controlled by a ____________.

54. Protein synthesis within cells is a complex process. A cytoplasmic organelle called the ribosome is required, as are several enzymes and a number of small molecules. In light of this information, would you predict that lysozyme would form if you mixed the amino acids that are present in lysozyme in a test tube? Explain your answer.

Polypeptide chains fold, coil, and bend. Each polypeptide or protein develops a shape that is unique to that protein and essential for its normal function. The shape of a particular protein is determined by its primary structure. Alterations of pH, temperature, or solvent around a protein may cause it to unfold, thus destroying its normal shape. Protein unfolding is called denaturation and destroys the function of the protein.

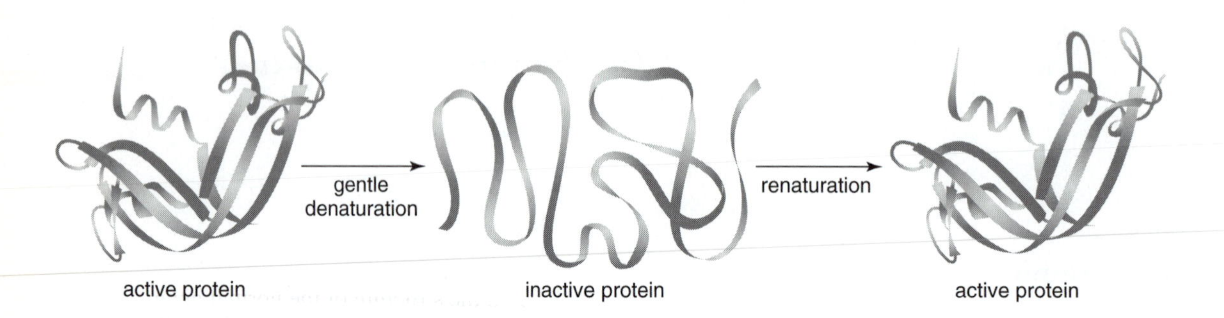

Questions

55. Hemoglobin is a carrier protein that transports ______ within the bodies of some animals. Hemoglobin has a unique shape. If you raise the temperature or change the pH extensively, the hemoglobin will ____________, a process called denaturation. Predict whether denatured hemoglobin can transport oxygen, and explain the reasons for your answer.

56. Raw egg white contains substantial amounts of a protein called egg albumin in colloidal dispersion. Describe the appearance of raw egg white. ____
An egg can be cooked by raising the temperature to 100° C for a few minutes. Describe the appearance of the egg white in a hard-boiled egg. ____

57. A temperature of 100° C causes egg albumin to unfold, a process called ____________. What easily detected physical change indicates that a protein has been denatured?

58. The shape of hemoglobin and the shape of egg albumin would be ____________ (the same or different), because each protein has a shape that is unique.

59. The ___________ of normal hemoglobin is controlled by the primary structure of hemoglobin. The primary structure of hemoglobin is controlled, or specified, by the ___________ for hemoglobin.

60. Sickle cell anemia (SCA) is a genetic disease that is inherited. People with SCA have hemoglobin that has a change in one amino acid, as compared to normal hemoglobin. The ___________ structure of SCA hemoglobin is different from normal hemoglobin. The shape of SCA hemoglobin is also a little different from normal hemoglobin, because the shape of a protein is determined by _______________.

61. The primary structure of a protein is determined by ___________. People with SCA have inherited a different version of a gene for hemoglobin, as compared to people who do not have SCA. A permanent, heritable change in a gene is called a mutation. An ancestor of a person with SCA must have had ___________ in a hemoglobin gene that altered the structure of the normal gene.

A polypeptide chain folds into its active shape in a series of stages. The first level of folding or coiling produces the secondary structure. The α helix and β pleated sheet are common secondary structures that are stabilized by the formation of many hydrogen bonds in the polypeptide backbone. The next phase of folding and bending forms the tertiary structure, producing protein shapes that may be compact and globular or fibrous and stringy. Interactions of the amino acid R groups with each other and with water stabilize the tertiary level of structure. The quaternary level of structure is present when two or more polypeptide chains associate to form a functional protein. The diagrams on the facing page illustrate these levels of structure.

Questions

62. An α helix is an example of a ____________ structure. Certain sequences of amino acids coil into an α helix. Many ____________ bonds form between parts of the polypeptide backbone that are closely positioned by the coiling. Hydrogen bonds are ____________ (weak or strong) bonds.

63. The β pleated sheet is a type of ____________ structure that is stabilized by the formation of many ____________. Certain sequences of amino acids fit well into the β pleated sheet pattern. In the case of the α helix, the many hydrogen bonds that stabilize the structure form between nearby regions of one polypeptide backbone. How does the pattern of hydrogen bonding differ in the β pleated sheet?

64. The variable groups, or R groups, of the amino acids may be nonpolar, polar, or ionized. Nonpolar R groups tend to be ____________ (hydrophilic or hydrophobic). Predict whether these R groups are likely to end up folded to the inside of the protein or on the outer surface. Explain your prediction.

65. A positively charged R group and a negatively charged R group could form an ____________ bond if bending of the polypeptide chain places them close together.

66. The most plentiful compound in living tissue, and the solvent of life, is ____________. Polar R groups can form ____________ bonds with water if they are close to water molecules. Based on this, predict whether polar R groups will tend to be folded to the inside or outside of the protein, and explain your reasons.

67. A protein that contains just one folded polypeptide chain exhibits the ____________, ____________, and ____________ levels of structure. The quaternary level of structure is present only in those proteins that contain ________________ polypeptide chain(s).

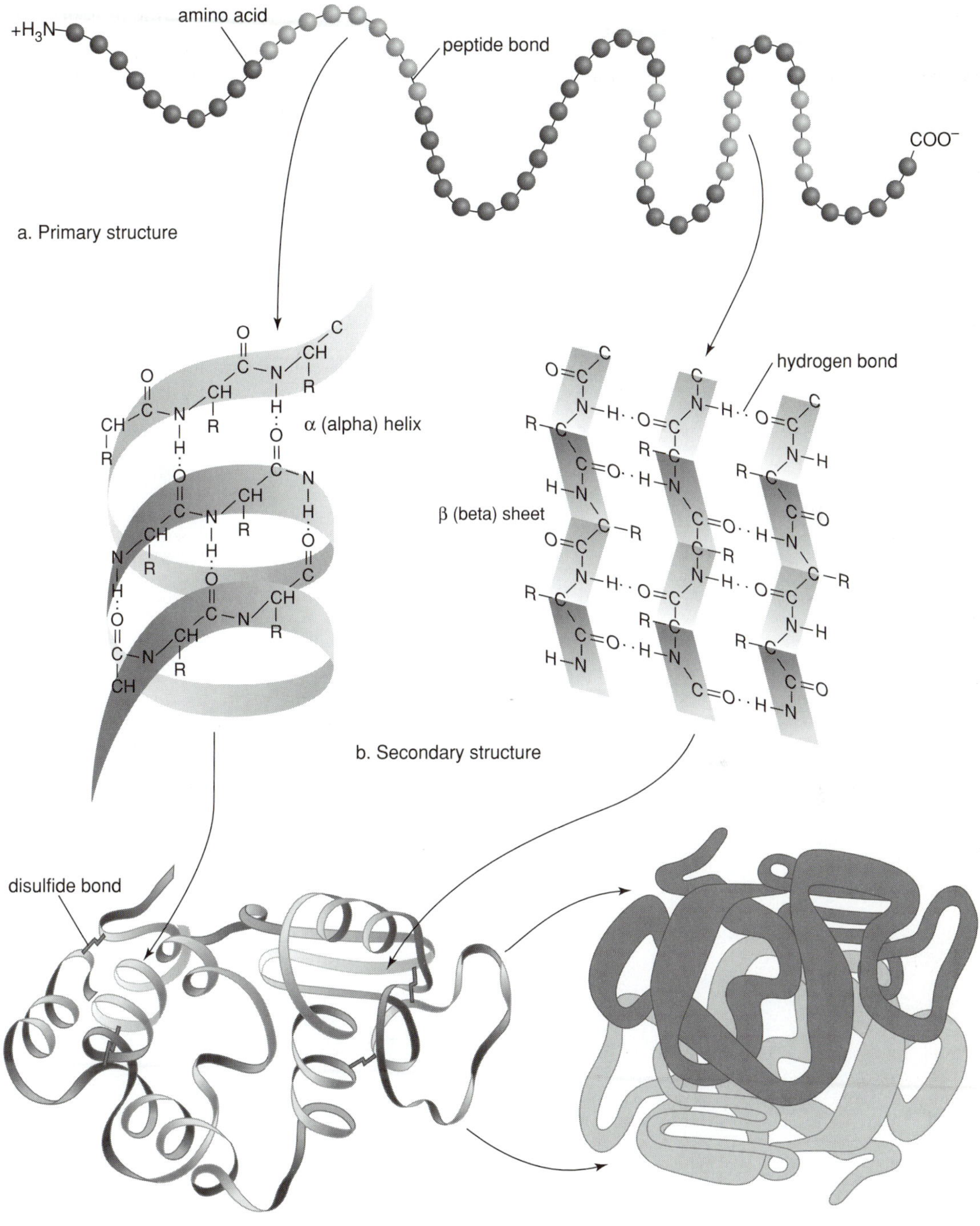

^+H_3N
amino acid
peptide bond
COO^-
a. Primary structure
α (alpha) helix
hydrogen bond
β (beta) sheet
b. Secondary structure
disulfide bond
c. Tertiary structure
d. Quaternary structure

68. Denaturation of a protein can be caused by moderate changes in the environment of the protein. Denaturation unfolds the protein and destroys the ____________, ____________, and ____________ levels of structure, but not the ________________. After denaturation, the protein ______ function normally. What kinds of bonds are broken when a protein is denatured?

69. Digestion of the primary structure of a protein breaks the ____________ bonds that link amino acids together and is called ____________. These bonds are stable, so the hydrolysis reaction requires either extreme chemical conditions or an enzyme called a ____________ to bring it about.

Each folded protein has a unique three-dimensional shape. Other molecules that have a matching shape will be able to interact with the protein in a specific manner. Complementary shapes are those that fit together like jigsaw puzzle pieces.

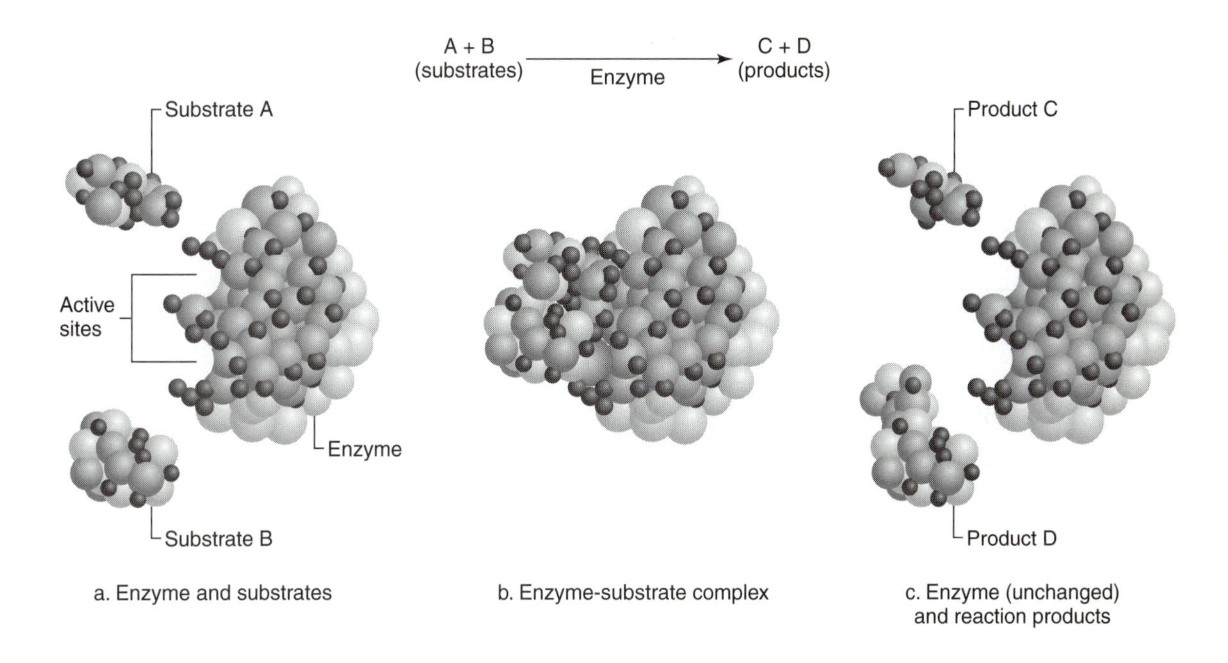

a. Enzyme and substrates

b. Enzyme-substrate complex

c. Enzyme (unchanged) and reaction products

Questions

70. An enzyme is a protein that functions as a ____________. The reactants for an enzyme-catalyzed reaction are called ____________.

71. Enzymes are highly specific, because they have one or more pockets called ____________ that are ____________ in shape to the substrates. Molecules other than the substrates will have different shapes and ____________ bind to the enzyme.

72. Denaturation of an enzyme or other protein causes it to ____________. After denaturation, are the active sites still present? Explain your answer.

73. A denatured enzyme ____________ function as a catalyst. Explain your answer.

74. Some proteins are embedded in membranes, and they transport substances through the membrane. The glucose transporter is a protein that facilitates the movement of glucose into cells. The glucose transporter does not transport fructose or other similar sugars. This suggests that there is a binding site on the glucose transporter that is ____________ in shape to glucose. Fructose and other sugars have shapes that ____________ the same as glucose and ____________ bind to the glucose transporter.

Proteins are flexible molecules that move as they perform their normal functions. A substance that binds to a protein is called a ligand. Ligand binding (or dissociation) often triggers a slight movement of the protein. The reshaping of a protein in response to binding by a ligand is called a conformation change. Conformation changes are essential to normal protein function.

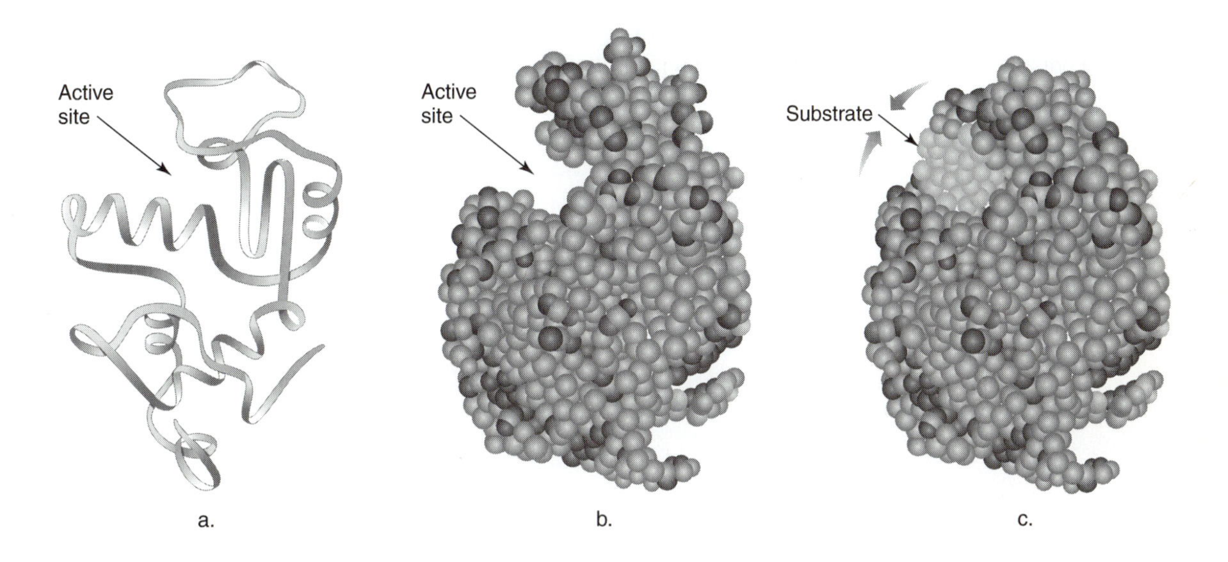

Questions

75. Substrates bind to enzymes at a specific location called the ____________ that is complementary in shape to the substrates. Enzymes are very specific, because the substrate and ____________ must ____________.

76. Anything that binds to a protein is called a ____________. The ____________ that bind at the active sites of enzymes are examples of ____________.

77. Binding of a ligand to a protein usually causes a __________________, a slight reshaping of the protein. Binding of a substrate at an active site triggers a __________________ in an enzyme, frequently causing the enzyme to grip the substrate(s) tightly.

78. During a chemical reaction, bonds must be broken in reactants before new bonds can form in the products. When an enzyme undergoes a conformation change in response to substrate binding, it may bend the substrate and strain the chemical bonds within it. This will make it ____________ to break bonds in the substrates, promoting formation of products at a ____________ rate.

79. Glucose binds to the glucose transporter and is therefore a ____________. Binding of glucose to the glucose transporter can trigger a __________________ that aids in the movement of the glucose to the interior of the cell.

80. A conformation change is a slight reshaping of a protein, triggered by ____________. Denaturation is a complete ____________ of the protein that destroys its normal shape and ____________. Denaturation is caused by __________________.

81. Hydrolysis of a protein requires either a ____________ (a protein-digesting enzyme), or extreme chemical conditions. Hydrolysis breaks all the ____________ bonds in the protein. After hydrolysis, the protein no longer exists. Instead, there is now a mixture of all the ____________ that were present in the protein.

Carbohydrates

Sugars, starches, and cellulose are all carbohydrates. Some sugars are important cellular fuels that are metabolized for energy. Simple sugars contain the atoms C:H:O in a ratio of 1:2:1. Starch and cellulose are polymers that are formed by linking many glucose units. Starch is used as an energy reserve, whereas cellulose is a plant fiber that is used in cell walls.

Questions

82. Glucose, fructose, and ribose are all simple sugars. They contain the elements ____________ in a ratio of ____________.

83. Glucose and fructose each contain six carbons. They also contain ______ hydrogens and ______. Glucose and fructose have the same chemical formula but different structural formulas. Glucose and fructose are ____________ of each other.

84. Ribose contains five carbon atoms, ______ hydrogen atoms, ______ oxygen atoms, and ______ nitrogen atoms.

85. Glucose, fructose, and ribose are sometimes called monosaccharides. The prefix mono- means ______. Saccharide means ____________.

86. Sucrose is a disaccharide that contains a glucose linked covalently to a fructose. Disaccharide is a term that means ____________.

87. A polysaccharide would be a macromolecule composed of ______ (many or a few) linked sugar units. Starch and cellulose are polysaccharides. The monomer that is polymerized to form starch and cellulose is ______.

88. Glucose is an important sugar that is often used as a fuel to provide ____________ to cells. Glucose is also useful because many other biochemicals can be synthesized using glucose as a starting material. Glucose ______ (is or is not) a valuable sugar for the cell.

89. Sugars ______ soluble in water. Honey and pancake syrup are concentrated sugar solutions. Describe the properties of concentrated sugar solutions.

90. Glucose is a molecule that is valuable to the cell. However, accumulating too much sugar can cause problems. Converting sugar to starch, a ____________, forms a molecule that is more suitable for storage. Starch is stored in starch grains in some cells. If sugar is needed by the cell, the starch can be ____________ to release the glucose that is needed.

The structures of several monosaccharides are shown in the diagram below. Each sugar contains a carbon backbone, several alcohol groups, and either an aldehyde or a ketone group.

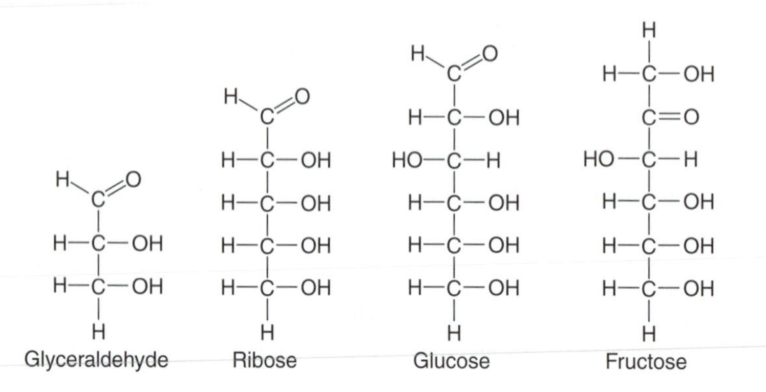

Questions

91. Glyceraldehyde is a triose that contains ______ carbon atoms in its backbone. Ribose is a pentose that contains ______ carbon atoms in its backbone, and glucose and fructose are hexoses that contain ______ carbon atoms in each sugar molecule.

92. The suffix -ose means ______. The numerical prefixes that are used indicate how many ____________ are present in the sugar.

93. There are ______ alcohol groups in glyceraldehyde, ______ alcohol groups in ribose, and ______ alcohol groups in glucose and fructose. What general rule can be used to determine the number of alcohol groups in a sugar?

94. Every sugar contains a carbonyl group. If the carbonyl group is on the end of the molecule, it is called a(n) ____________. Carbonyl groups that are located between other carbon atoms are called ____________.

95. The functional groups that are present in sugars are ____________. Because of this, sugars are hydrophilic and ____________ dissolve in water.

96. Which sugar(s) contain an aldehyde group? ________________________ Which sugar(s) contain a ketone group? ________________________

97. A molecule contains large amounts of carbon and hydrogen, and small amounts of oxygen, nitrogen, and phosphorus. Would you classify this compound as a carbohydrate? Explain your answer.

Glucose that is dissolved in water usually forms a ring. When a glucose molecule closes into a ring, two configurations are possible, as shown in the diagram below. The ring forms of glucose are often represented with abbreviated structures. Each form of glucose is a monomer that can polymerize. Polymerization of α glucose forms starch, whereas polymerization of β glucose forms cellulose.

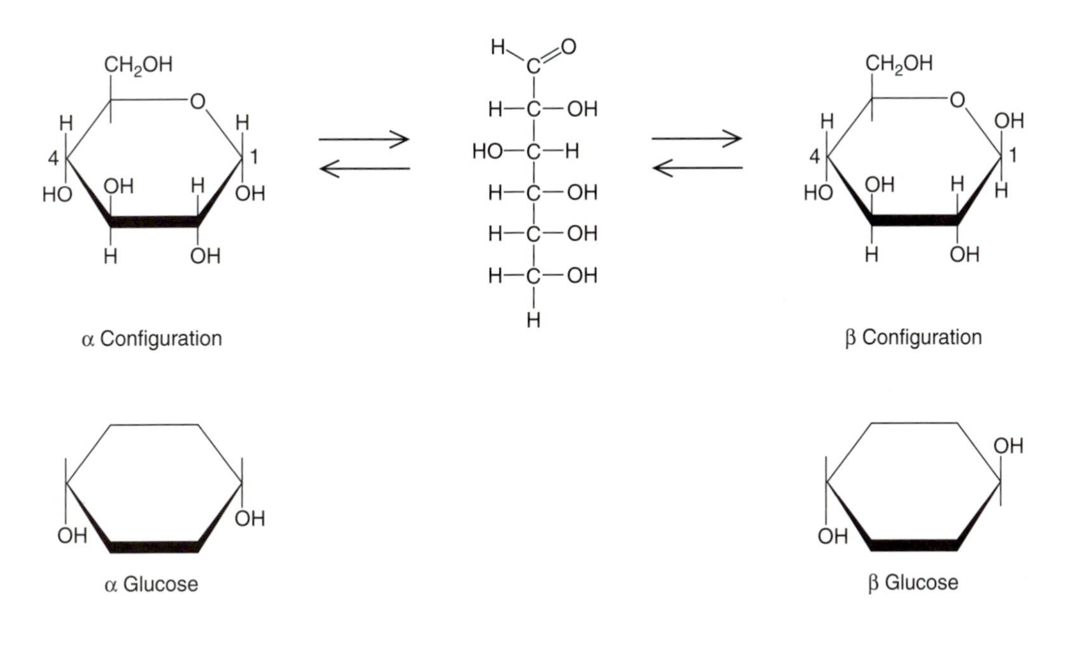

Questions

98. Most dissolved sugar molecules form a ____________. When glucose closes into a ring, two configurations are possible, called __________________.

99. Compare the structures of α glucose and β glucose. These forms of glucose are similar, because in both cases, the glucose molecule has closed on itself to form a ____________. How do α glucose and β glucose differ from each other?

100. The shapes of α glucose and β glucose ____________ (are or are not) slightly different. The ______ form of glucose can polymerize and form starch. Polymerization involves a reaction called __________________, which is catalyzed by a specific enzyme. The enzyme that catalyzes the synthesis of starch has an active site that is complementary to ______ glucose, but not to ______ glucose.

101. The ______ form of glucose can polymerize and form cellulose. Polymerization involves a reaction called __________________, which is catalyzed by a specific enzyme. The enzyme that catalyzes the synthesis of cellulose has an active site that is complementary to ______ glucose, but not to ______ glucose.

102. The hydrolysis of starch and water to form ____________ molecules is the principal reaction of starch digestion. Starch and water are stable and do not react unless an appropriate ____________ is present to catalyze the reaction. Hydrolysis of starch releases ______ molecules that can then be used as an energy source, or fuel, by the cell.

103. When cells contain excess sugar, they typically polymerize it with the aid of enzymes, and store the product ____________, until sugar levels in the cell are low. When the cell needs sugar again, it can ____________ some of the stored starch and replenish its supply of ____________. Describe the biological role of starch.

The diagram below shows a dehydration synthesis reaction between one glucose molecule and one fructose molecule. The new bond that links the sugars together to form sucrose is called a glycosidic linkage.

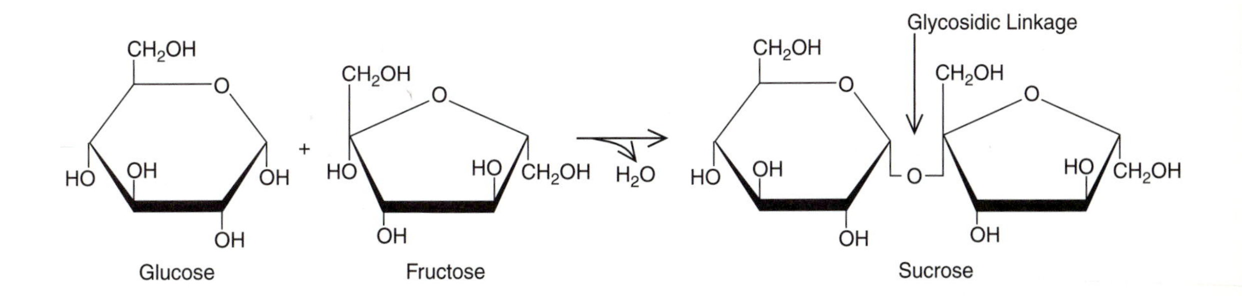

Questions

104. When you purchase "sugar" in the grocery store, the bag contains sucrose. Sucrose contains a molecule of glucose linked to a molecule of ____________ by a ____________ linkage. Because it contains two linked sugars, sucrose is a ____________.

105. When dissolved glucose and fructose are mixed, the ____________ reaction to form sucrose does not happen at detectable rates. What is required to speed this chemical reaction along?

106. Digestion of sucrose to form ____________ (what sugar) and ____________ (what sugar) occurs by a reaction called ____________ that also uses a molecule of ____________. Dissolved sucrose is stable. What is required to speed this chemical reaction along?

107. Hydrolysis of lactose, a sugar found in milk, releases galactose and glucose. Because it contains two linked sugars, lactose is a ________________. Lactase is an ____________ composed of ____________ that speeds the hydrolysis of lactose. Hydrolysis of maltose releases two molecules of glucose. Maltose is a ____________. The hydrolysis of maltose is catalyzed by an enzyme called ____________.

108. Starch is a ____________ that forms when many α glucose units are linked together by ____________ linkages. Starch functions as an energy reserve, because it can be hydrolyzed to release ____________ as the cell or organism needs it.

109. Cellulose is a ____________ that forms by linking many β glucose units together. Cellulose is made by plant cells and secreted to the outside surface of the cell, where it forms a major component of plant cell walls. Cellulose is used as a structural molecule by plant cells and does not function as a ____________ like starch does.

110. Cellulase is an enzyme that can hydrolyze cellulose and release ____________. Humans do not synthesize cellulases and cannot ____________ cellulose. Cellulose is a polymer and is therefore a macromolecule. Macromolecules ____________ (can or cannot) be absorbed and used by the body. Is cellulose a good calorie source for humans? Explain.

The diagrams below represent partial structures of several types of polysaccharides, all glucose homopolymers. Since the various forms of starch and cellulose are macromolecules, the complete molecules are much larger than the partial structures shown below.

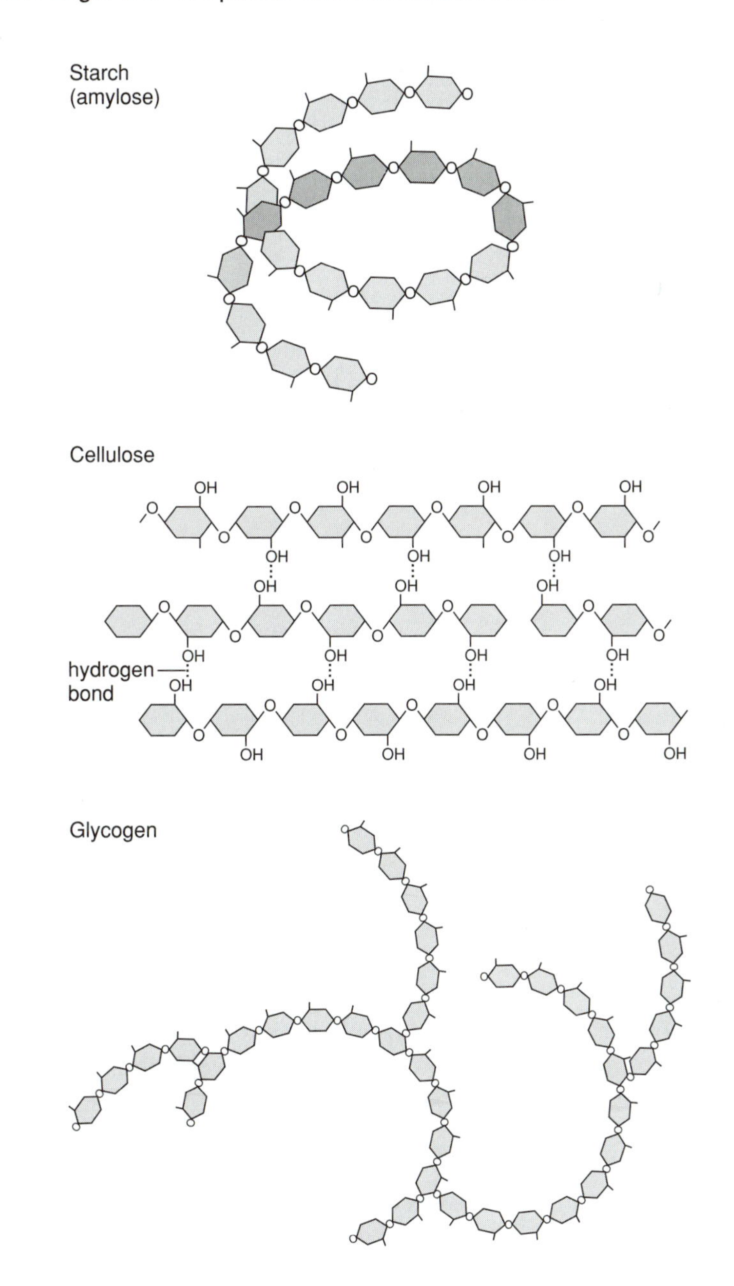

Questions

111. Hydrolysis of glycogen releases ____________. Hydrolysis of amylose (a plant starch) releases ____________, and hydrolysis of cellulose releases ____________. All these polysaccharides release only ____________ when they are hydrolyzed and are therefore called __________________. (heteropolymers or homopolymers)

112. Homopolymers exhibit less diversity in their structures, compared to ____________. Homopolymers such as the polysaccharides have ____________ (more or fewer) biological functions, because ______________________.

113. Glycogen is an energy-storing polysaccharide that is synthesized and stored in some animal tissues. Amylose is an energy-storing polysaccharide that is synthesized and stored in some plants. These molecules are polymers of ____________, and all can be hydrolyzed to release ____________ when it is needed.

114. How do the structures of glycogen and amylose differ?

115. Cellulose is a polysaccharide that is used as a structural material by many ____________ cells and is a major component of ____________. How does the structure of cellulose differ from the polysaccharides that function as fuel (energy) reserves?

116. Organisms such as cows and termites can live on a diet that contains large amounts of cellulose (from plants and wood). They contain microorganisms in their digestive tracts that aid digestion, because they have enzymes called ____________ that can ____________ cellulose.

117. Cellulose is a form of dietary fiber that passes through the digestive tracts of many animals, including humans, without undergoing hydrolysis. Explain how these animals differ from termites and cows.

Lipids

Lipids include a variety of compounds that are fatty, oily, or waxy. Lipids are composed mostly of hydrocarbon and are hydrophobic. Various lipids are listed below, along with their major biological functions. The first three groups will be discussed in later sections.

Lipid	Example and Description
Fats and oils	Fats and oils are a major energy reserve in animals and in plant seeds.
Phospholipids	Modified fats that form bilayers—the basis of all biological membranes.
Steroids	Various steroids are hormones, structural materials, vitamins, and bile salts.
Prostaglandins	Local communication and roles in inflammation, pain, and smooth muscle contraction.
Waxes	Prevent excessive water loss from leaves.

Questions

118. Lipids are not very soluble in water, because they are composed mostly of ____________ and are nonpolar. Structures of this type are described as ____________.

119. Hydrocarbons have a high content of stored ____________ and can be burned or oxidized to release that energy. Some lipids are used as energy reserves by cells, because they store a lot of ____________ (measured as calories) in relatively little weight.

120. Cholesterol is a lipid that is a component of membranes in animal cells. Cholesterol functions as a ____________ component of membranes.

121. Terrestrial plants need to conserve water. Their leaves are often coated with waxy materials that ______ permeable to water.

122. Prostaglandins are composed mostly of ____________ and are therefore ______ soluble in water. Prostaglandins play roles in intercellular communication. They are lipids that play a role in many processes such as inflammation, ____________, and ____________.

123. Adipose or fat tissue is a specialized tissue that stores __________ in the bodies of many animals. Because fats are composed mostly of __________, they store a great deal of energy that can be released when the fats are oxidized. Fats are good fuels, because they store large amounts of __________, yet are low in density because of the high content of hydrogen, an element with an atomic weight of about _____.

124. Because most animals are mobile, it is an advantage to store as much energy (as many calories) as possible in the least weight possible. Explain why fats are the best biomolecules for this purpose.

Fats and oils are also called neutral fats or triglycerides. Each fat molecule contains three fatty acids linked to one glycerol molecule. The structure of the glycerol molecule and examples of fatty acids are shown in the diagrams below.

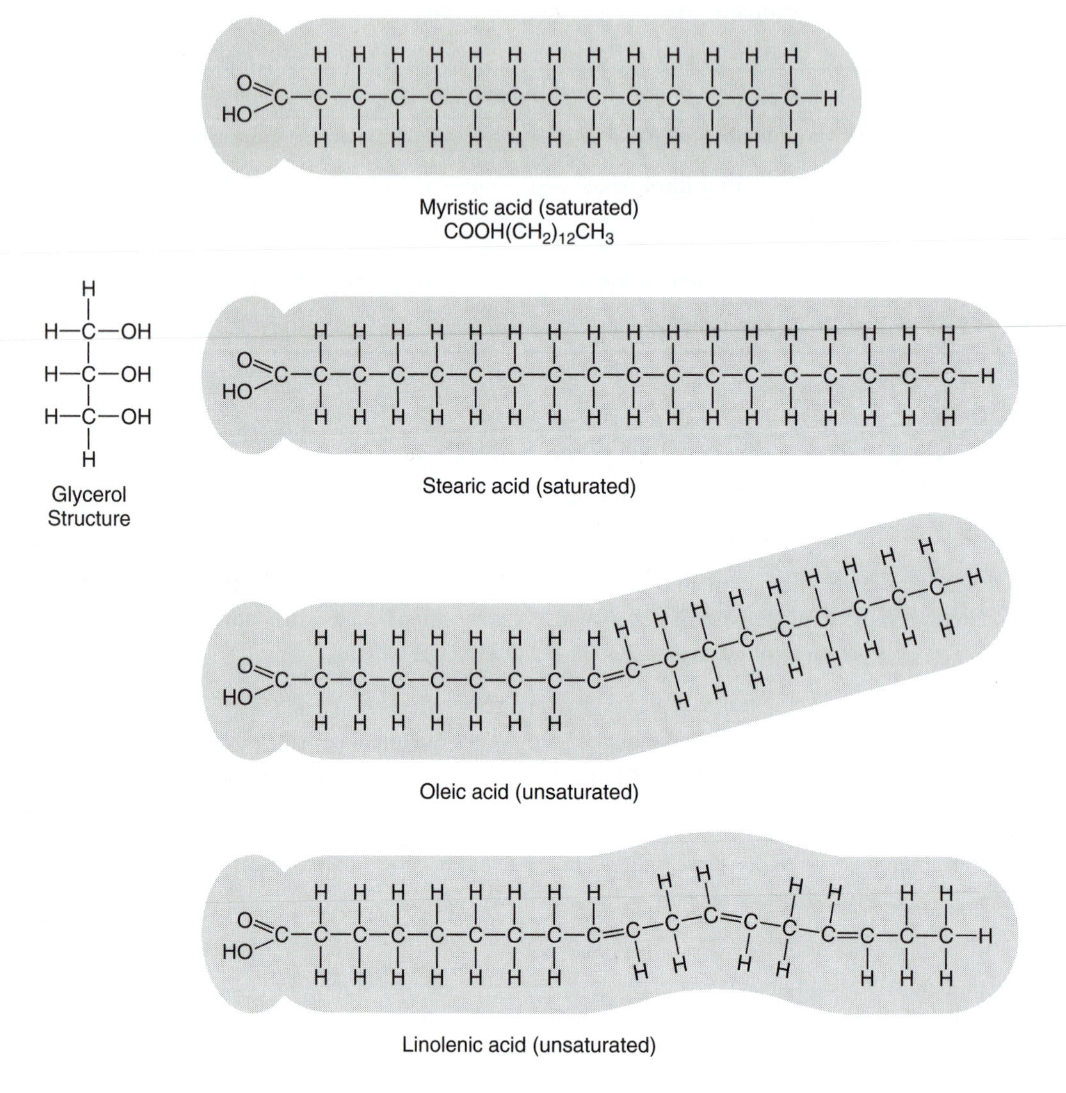

Questions

125. Glycerol is a molecule that contains ______ carbon atoms, ______ hydrogen atoms, and ______ oxygen atoms. What functional group(s) is present in the glycerol molecule? ______.

126. Compare the structures of all the fatty acids to answer the following questions. All fatty acids contain a long chain that is composed of ____________ and have a ____________ group at one end of the molecule. The hydrocarbon part of a fatty acid is nonpolar and therefore ____________. The carboxyl group is so polar that it is partially ionized and releases ____________ ions into solution.

127. In light of your answers to the previous questions, explain why "fatty acid" is a good name for molecules of this type.

128. The nonionized form of the first fatty acid is called myristic acid. The ionized form is called myristate and has a negative charge on the carboxyl group, because a ____________ ion has dissociated. The ionized form of stearic acid would be called ____________. After a hydrogen ion dissociates from the oleic acid, the structure that remains is called ____________.

129. Count the carbon atoms that form the backbone of each fatty acid. Myristic acid contains ______ carbons, oleic acid contains ______ carbons, linolenic acid contains ______ carbons, and stearic acid contains ______ carbons. All the fatty acids shown here contain ______ (even or odd) numbers of carbon atoms.

130. Abbreviated formulas for fatty acids, such as the example shown for a myristic acid, do not show the positions of all the bonds. Write abbreviated formulas of the same type for the other three fatty acids.

131. Compare the hydrocarbon chains of the fatty acids to answer these questions. In the saturated fatty acids, all the carbons are linked to each other by ____________. All the carbons except those at the ends of molecules are also bonded to ____________ (how many) hydrogens. In the unsaturated fatty acids, most of the carbons are linked to each other by ____________, and are also bonded to ______ hydrogen atoms. Some carbons are bonded to each other by a ____________, and these carbons are bonded to only one hydrogen.

132. Which type of fatty acid contains the highest proportion of hydrogen, the unsaturated group or the saturated group? __________________. Saturated fatty acids are bonded to the maximum amount of hydrogen, because all the carbon-to-carbon bonds are ____________. Saturated fatty acids are saturated with ____________.

133. Unsaturated fatty acids contain ______ hydrogen than a comparable saturated fatty acid. Some of the carbon-to-carbon bonds are ____________, and because carbon can form ______ (how many) bonds, it has only ______ valence position available and therefore bonds to one hydrogen.

134. Compare the shapes of the unsaturated fatty acids to the saturated fatty acids.

135. A reaction called hydrogenation would add ____________ to a molecule, and a reaction called __________________ would remove hydrogen.

136. An unsaturated fatty acid can be converted to a saturated fatty acid in a reaction called __________________, which breaks double bonds and adds ____________ to the structure.

137. A saturated fatty acid can be converted to an unsaturated fatty acid in a reaction called ____________, which removes ____________ from the structure and leads to the formation of ____________ bonds between some carbon atoms. Dehydrogenation is a type of oxidaton reaction.

138. Oleic acid could be converted to stearic acid by the reaction of ____________. Since removing hydrogen atoms is a way of oxidizing a molecule, adding hydrogens is a way of carrying out the process of ____________.

The reaction of dehydration synthesis can link three fatty acids and one glycerol to form a fat and three water molecules. Fatty acids are then attached to the glycerol by ester linkages.

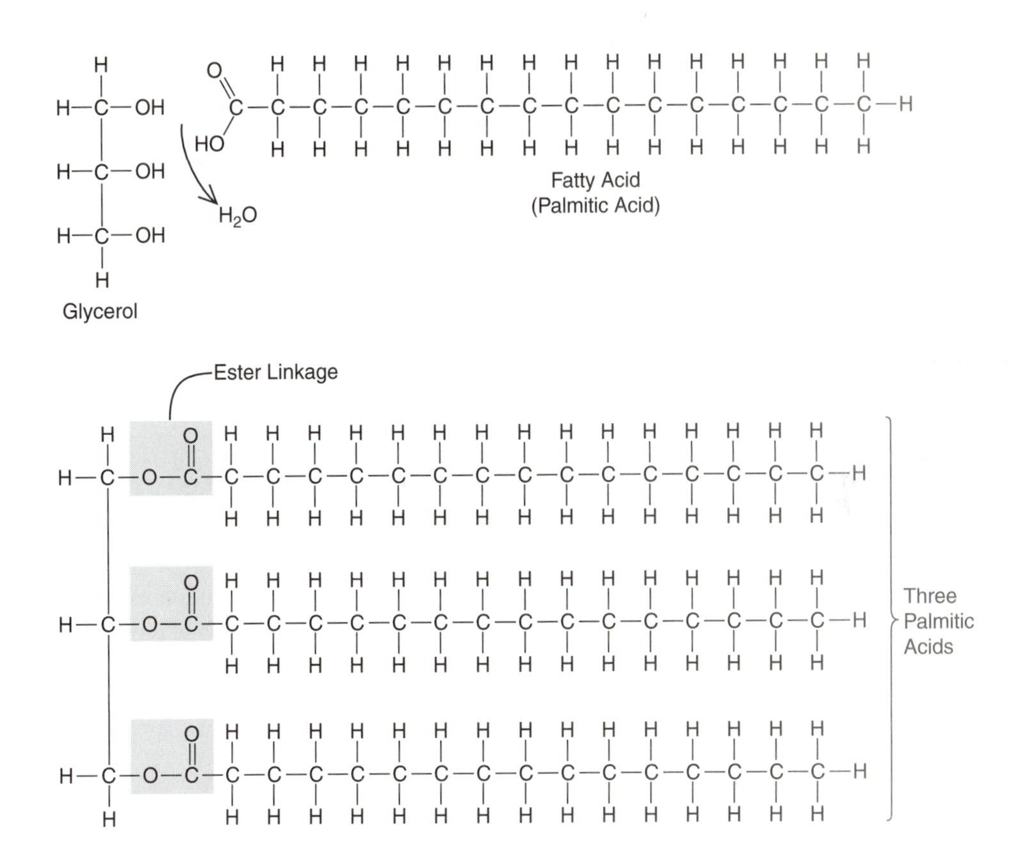

Questions

139. A fat contains ______ (how many) fatty acids linked to ______ glycerol molecule with ______ ester linkages.

140. Ester linkages can be broken during the reaction of ____________. If all three ester linkages in a fat molecule are broken, the products are ____________ and ____________.

141. Fat synthesis occurs in an animal when it consumes more food than it needs at that time. The reaction of ____________ to form fats predominates when food is plentiful and animals eat extra amounts. Fat can be stored in the body of the animal in adipose tissue. Adipose tissue tends to ____________ when an animal is overfed.

142. Hungry animals that do not find food immediately can obtain fuel by ____________ some of their body fat to release glycerol and ____________, which can be oxidized and used as ____________.

143. The biological role of fats is to serve as __________________.

144. Unsaturated fats contain one or more ____________ fatty acids, whereas saturated fats contain ____________ fatty acids.

145. Saturated fats contain fatty acids that ______ have double bonds and ______ saturated with hydrogen. Saturated fatty acids are solid at room temperature. Give some examples of fats that are mostly saturated, based on this characteristic.

146. Unsaturated fats contain at least one fatty acid that ____________ (does or does not) have double bonds and ____________ saturated with hydrogen. Unsaturated fats are liquid at room temperature and are often called oils. Give some examples of unsaturated fats.

147. Fats are also called neutral fats and triglycerides. Explain why these are good names for the molecule pictured on the previous page.

148.*The shape of a saturated fatty acid is ____________, whereas the shape of an unsaturated fatty acid is ____________. Explain how these shape differences could have an effect on the melting point of fats.

149. Some evidence suggests that excessive consumption of saturated fats contributes to the onset of blood vessel disease in humans. Which fats should be minimized in the diet to avoid this problem?

150. If you do not have a chemical analysis of a fat, how can you easily tell if it is saturated?

151. Corn oil is ____________ at room temperature. Corn oil contains mainly ____________ fats. Margarines are ____________ at room temperature. An ingredient in many margarines is "partially hydrogenated corn oil." Explain what has happened chemically to the corn oil and why its melting point has changed.

152. An unsaturated fat that is hydrogenated becomes a ____________ fat with a ____________ melting point.

153. Corn oil, sesame oil, olive oil, and peanut oil are all derived from ____________ and are all ____________. Most plants do not store much fat, except in their ____________. Explain why fats (or oils) are important components of many plant seeds. (Hint: Plant seeds contain an embryonic plant in addition to stored food.)

Phospholipids are modified fats. They are diglycerides that have polar groups such as phosphate and nitrogenous bases substituted for one of the fatty acids. A typical phospholipid is shown below.

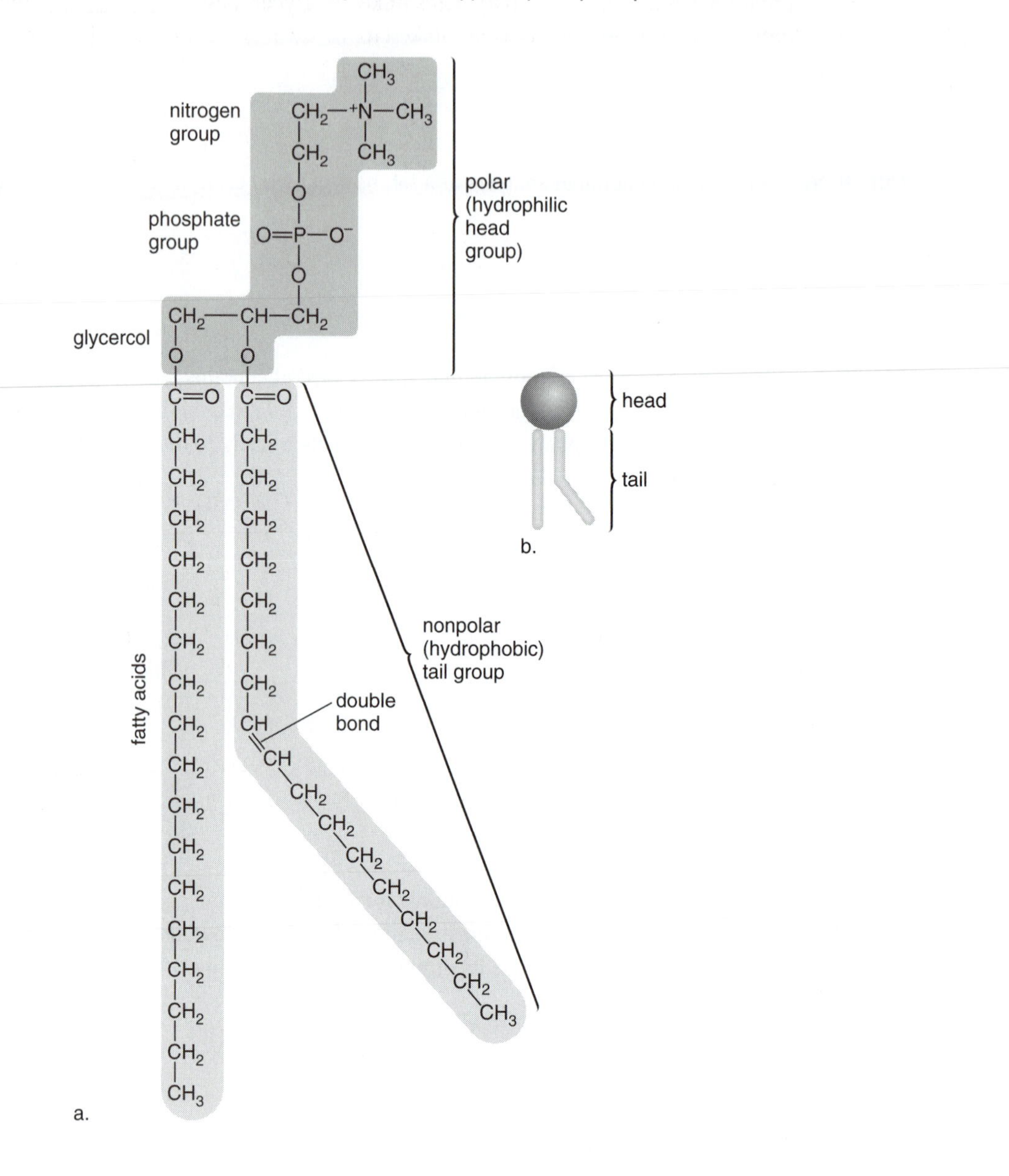

Questions

154. A phosphoplipid contains ______ fatty acids linked to glycerol by ____________ linkages. The third alcohol group on the glycerol is not linked to a fatty acid, but instead is bonded to an acidic group called ____________, which is derived from phosphoric acid. The phosphate is also bonded to a nitrogenous base. Organic compounds that contain nitrogen are weak bases, because they accept and bond to ______.

155. Part of the phospholipid is hydrophobic, because it is composed of ____________ and is ____________. Part of the phospholipid is ____________, because it contains groups that ionize.

156. The simplified phospholipid structure represents the hydrophilic region of the molecule with a ____________. This part of the phospholipid is ionized due to the presence of ________________.

157. The simplified phospholipid structure represents the nonpolar, hydrophobic portion of the molecule with ____________. This part of the phospholipid contains __________________.

158. One end of a phospholipid molecule is very ____________, and the other end is ____________ and hydrophobic.

159. When phospholipids are mixed with water, they tend to self assemble into structures called bilayers. Predict which part of the phospholipid molecule is likely to be in contact with water, and explain your prediction.

When phospholipids are in contact with water, they self assemble into a structure called a phospholipid bilayer. The hydrocarbon tails of the phospholipid are inside the bilayer, and the polar heads are outside and in contact with water. Phospholipid bilayers make up the basic structure of biological membranes. A small region of a typical phospholipid bilayer is shown below.

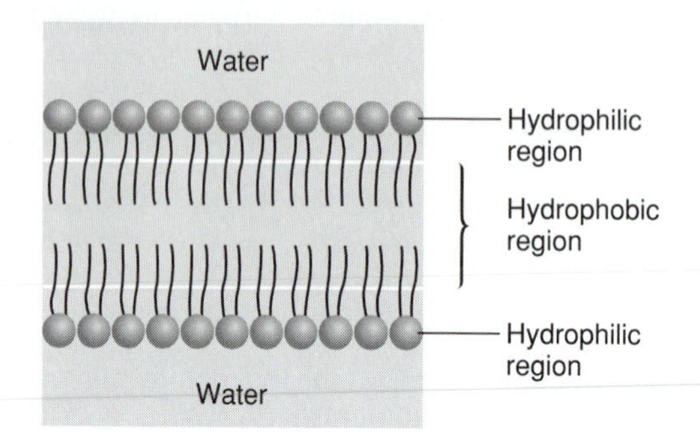

Questions

160. Phospholipid bilayers are composed of ______ layers of phospholipids, arranged so that the ____________ portion of the phospholipid is inside, and the ____________ portion of the molecule is outside where it can interact with water.

161. The phospholipid bilayer is constructed like a sandwich. The "filling" is composed of the portion of the phospholipid that is ____________, and the "bread" includes the portion of the phospholipid that is ______________.

162. The phospholipid bilayer self assembles. Explain why grouping all the hydrophobic tails on the inside is a stable arrangement.

163. Explain why grouping the hydrophilic heads next to each other and on the surface of the phospholipid bilayer is a stable arrangment.

164. Biological membranes are composed of ____________ as the basic structure. Phospholipid bilayers are used to separate one aqueous compartment from another, because they are stable structures that ____________ when phospholipids and water are mixed

165. The structure of biological membranes includes many proteins that are embedded in the basic structure of the ____________. Proteins in the membrane are important for transport of materials in and out of cells. Give one example of such a protein, and explain how it functions.

Steroids are classified as lipids, because they are composed mostly of hydrocarbon and are therefore hydrophobic. All steroids contain a core of four fused hydrocarbon rings. The structures of two steroids are shown in the diagram below.

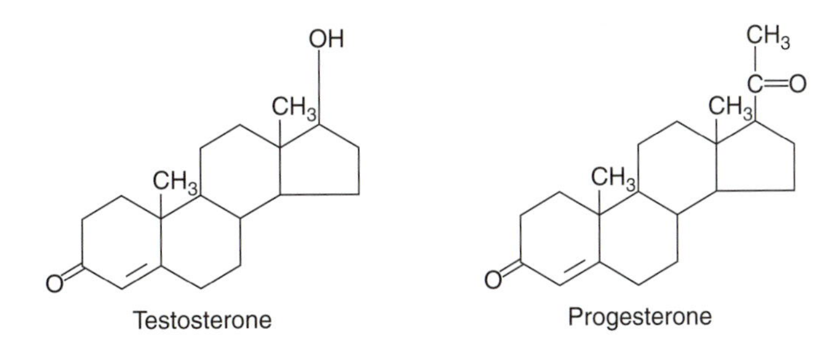

Questions

166. All steroids are similar because they contain ______ rings composed of ______. How do specific steroids differ from each other?

167. Cholesterol is a ____________ that is present in the membranes of animal cells. Cholesterol is made mostly of ____________, and is therefore ____________. Predict the specific location of cholesterol in the membrane, and explain the reasons for your prediction.

168. Testosterone is an example of a steroid that functions as a ____________. Hormones circulate through the body and bind to receptors made mostly of ____________ found in the membranes of target cells. Hormones produce changes in target cells after binding to them.

169. Vitamin D, bile salts, female sex hormones, and many adrenal hormones are steroids. All steroids are similar, because they contain ________________. How would each of these steroids differ from other steroids?

170. Steroids play a variety of roles in the body. ________________ is found in the membrane of animal cells and plays a structural role. Testosterone is an example of a ________________, a regulatory molecule that alters its target cells.

Nucleic Acids

Nucleic acids include two closely related classes of molecules, deoxyribonucleic acid (DNA) and ribonucleic acid (RNA). Both DNA and RNA are unbranched heteropolymers. Genes are the structures that store genetic information, and they are composed of DNA. RNA is involved in various aspects of gene expression.

Questions

171. Both ______ and ______ are nucleic acids. The subunits that polymerize to form DNA and RNA are called _________. Since DNA and RNA are heteropolymers, there is ___________ type of nucleotide.

172. DNA is a macromolecule that is often extremely large. Some DNA molecules are the largest biomolecules known. The molecular weight of a typical DNA molecule is ____________ the molecular weight of a typical RNA molecule.

173. Genes are composed of ____________. A closely related molecule, ____________, is important in various aspects of gene expression.

174. The primary structure of a protein is the sequence of amino acids in the protein and is specific and unique to each protein. The primary structure of a protein is controlled by a ____________ for that particular protein. Genes are composed of ____________. DNA is a heteropolymer that codes information that controls the __________________ of a particular polypeptide or protein.

175. There are four different nucleotides that can polymerize to form ______, the molecule that codes information in genes. There are no branches in DNA, so how do different DNA molecules differ from each other?

176. Different DNA molecules may have different sequences of ____________, and different proteins have different sequences of ____________. The information in genes is coded in ____________ molecules. The sequence of nucleotides in DNA ultimately controls the sequence of ____________ in proteins.

177. The sequence of amino acids in a protein determines how the protein coils and folds, and thus determines the ____________ of the protein. The shape of a protein is responsible for the ____________ of that protein. If the shape of a protein is destroyed by __________________, the protein no longer works. If the sequence of nucleotides in a gene is changed, the ____________ of amino acids in a polypeptide may change also. The shape and ____________ of the protein may then be affected. Give one example of a genetically altered protein. ________________________________

A nucleotide contains three linked groups: a phosphate, either deoxyribose (DNA) or ribose (RNA), and one of four nitrogenous bases. The diagram below shows these components and a typical nucleotide.

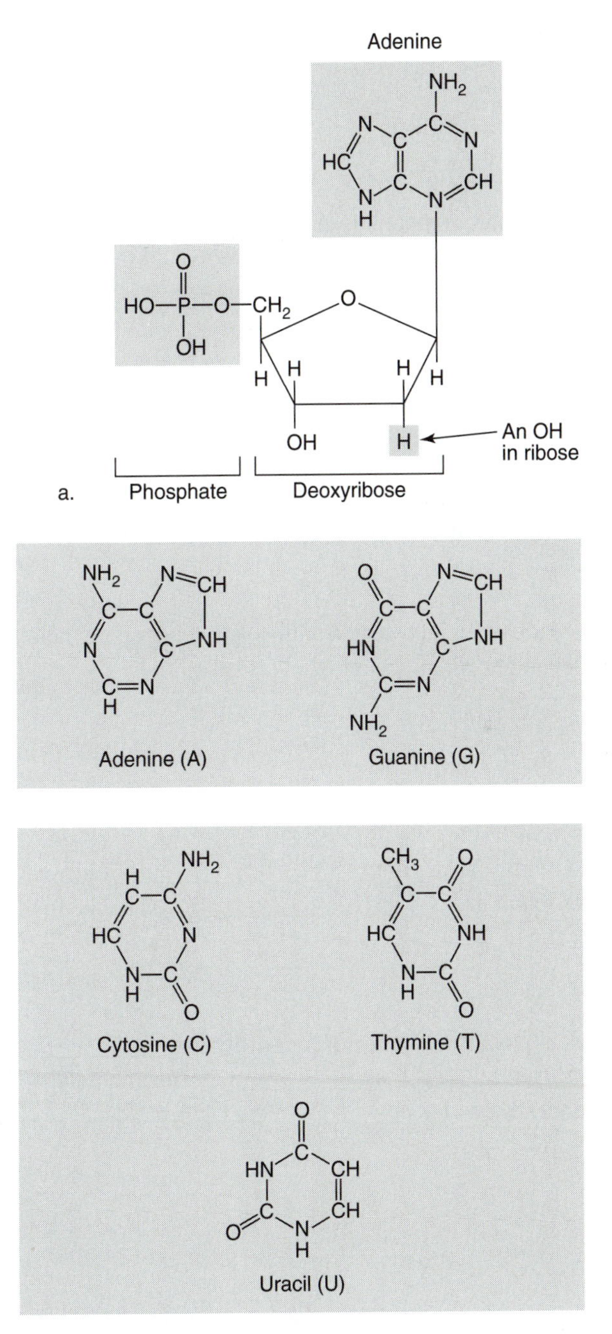

Questions

178. Nucleotides found in RNA always contain a pentose, a sugar containing ______ carbon atoms, called ____________. In contrast, DNA always contains ____________ rather than ribose. How does deoxyribose differ from ribose?

179. Nitrogenous bases are organic compounds that contain ____________. They function as weak organic bases, because they __________________. Which of the compounds on the previous page are nitrogenous bases?

180. Three of the nitrogenous bases ________________________ are found in both DNA and RNA. ____________ is found only in DNA, and ____________ is found only in RNA.

181. The nitrogenous bases are often represented with single-letter abbreviations. List the full name and the appropriate abbreviations for all the nitrogenous bases.

182. A nucleotide contains a phosphate group, __________________, and one of four nitrogenous bases. The four bases found in DNA include ____________, and the four bases found in RNA include ____________.

183. Different nucleotides in DNA contain one of four different ____________. A gene contains the information that controls the __________________ of a specific polypeptide. The information is coded in a unique sequence of ____________ in the DNA molecule.

184. Each gene codes for a unique ____________ and has a unique sequence of ____________.

Nucleotides are linked together by the reaction of dehydration synthesis. A short segment of a polynucleotide is shown in the diagram below. The phosphate ester bonds that link the nucleotides together can be broken by the reaction of hydrolysis.

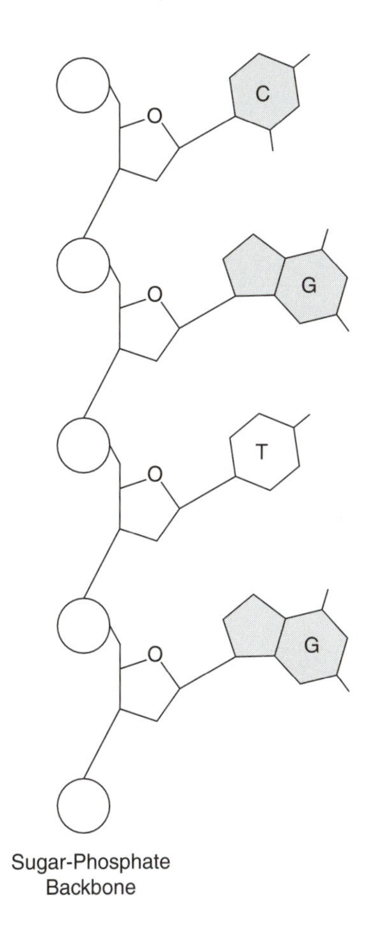

Questions

185. DNA and RNA are both ____________, formed by linking nucleotides together with __________________ reactions. Another product, ____________, is also formed during this reaction.

186. A polynucleotide has a backbone region that is always the same. The backbone always consists of alternating sugar and ____________ groups.

187. Where are the nitrogenous bases located in DNA and RNA? _________________________________

188. The sugar-phosphate backbone is the same in all DNA molecules, whereas the sequence of the ____________ varies.

189. All the bonds within a polynucleotide chain are ____________ and are therefore strong bonds.

190. When DNA is hydrolyzed, it releases a mixture of four different ____________. DNA is a stable molecule that does not undergo hydrolysis quickly when it is dissolved in water. What could you do to make this reaction run faster? ____________

191. Enzymes often have names that end in the suffix ____________. A nuclease is an enzyme that digests, or __________________, a nucleic acid.

192. An enzyme that digests ______ is called DNase, whereas an enzyme that digests RNA would be called ____________.

DNA is usually double stranded. Each of the two strands within DNA is a polynucleotide chain. The two strands of DNA are linked by many hydrogen bonds that form between the nitrogenous bases. A short fragment of DNA is sketched as a ladder in the diagram below. The actual shape of DNA is a double helix, a shape that forms when the base of the ladder is fixed in position and the top of the ladder is twisted.

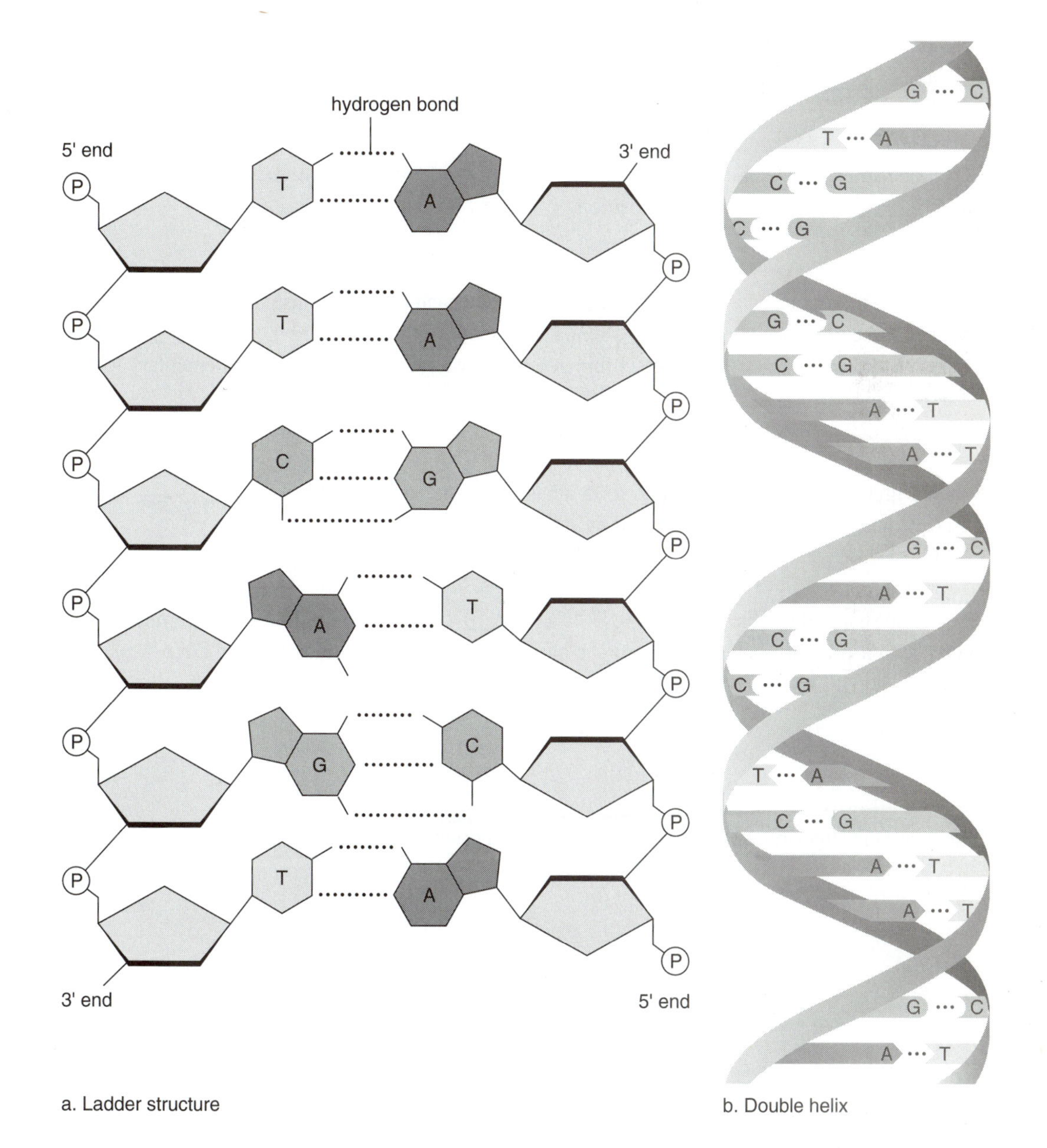

a. Ladder structure

b. Double helix

Questions

193. Within one strand of DNA, all the bonds are ____________ bonds that are strong and difficult to break. The bonds that link the two strands of DNA are all ____________ bonds, weak bonds that are easily broken.

194. When one strand of DNA contains a thymine (T) at a particular position, the second strand always has a ____________ at that same position.

195. When one strand of DNA contains a cytosine (C) at a particular position, the second strand always has a ____________ at that same position.

196. Particular nitrogenous bases are always found opposite each other, because they have shapes that fit together and can form ____________ bonds. The shapes of T and A are __________________. The shapes of C and ____________ are complementary, whereas the C is not complementary to the bases ____________.

197. When T and A are close together and form a base pair, they are linked by ____________ hydrogen bonds. G and C form a base pair linked by ____________ hydrogen bonds.

198. The actual shape of double-stranded DNA is a ____________. The double helix resembles a spiral staircase in its overall shape. Where are the base pairs located in double-stranded DNA?

Where are the sugar-phosphate backbones located in double-stranded DNA?

RNA differs from DNA in several ways. RNA contains ribose rather than deoxyribose, it contains the base uracil rather than thymine, and it is usually single stranded. A segment of RNA is shown in the diagram below. RNA molecules play roles in various aspects of gene expression.

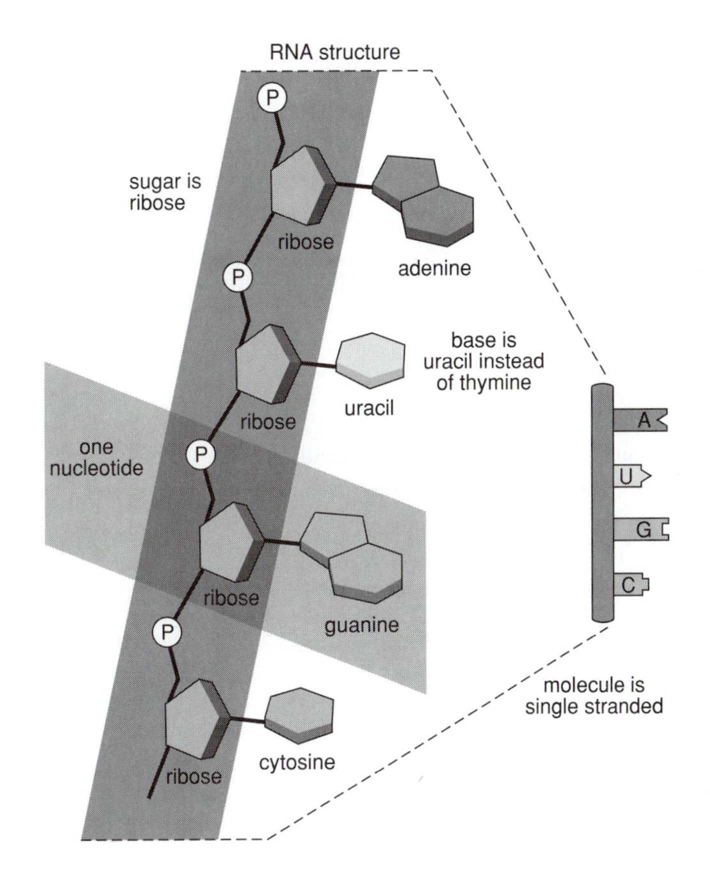

Questions

199. RNA is a ____________ chain that contains ____________ types of monomers linked together by phosphate ester linkages. Because more than one type of nucleotide is present in RNA, it is a ____________________.

200. DNA is double stranded, whereas RNA is usually ________________. List two other ways that RNA differs from DNA.

201. Doubled-stranded DNA molecules are fairly rigid, whereas RNA is ____________ and is much more flexible.

202. Some RNA molecules have double-stranded regions within them. Under some circumstances, RNA can bond temporarily to DNA. In a DNA–RNA hybrid, cytosine (C) pairs with ______, and an adenine (A) in DNA would be paired with a ____________ in RNA.

203. If a fragment of one DNA strand has the base sequence CCGTTACGA-, then the second strand of DNA would have the base sequence __________________. The base pairs in this fragment are connected by ____________ hydrogen bonds. The two strands can be separated only after the hydrogen bonds are ____________.

204. A gene is copied when a cell is ready to express that gene. The two strands of DNA must first separate, then one strand of DNA (the template strand) is copied. The copies are made of RNA and have a base sequence that is complementary to the DNA template. RNA that forms as a complementary copy of DNA with the sequence CCGTTACGA- would have the base sequence ____________________. How does this RNA sequence compare to the DNA sequence you derived in the last problem?

ATP is a molecule that functions as an energy shuttle within the cell. ATP stores chemical energy in its structure and can provide that energy to processes and reactions that need energy in the cell. The structure of ATP is related to one of the monomers present in RNA.

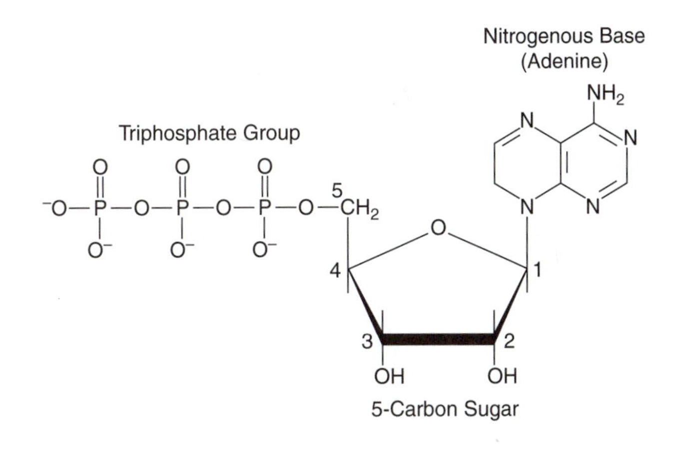

Questions

205. ATP stands for adenosine triphosphate. How is this molecule related to the monomers that are present in RNA?

206. Some chemical reactions require energy. These reactions do not proceed at a high rate unless they have an energy source. Within cells, such reactions obtain their energy from ____________.

207. Muscle contraction requires energy. The energy to power muscle contraction is provided by ______.

208. Pumping various solutes across membranes is called active transport. Active transport requires energy. The energy for active transport is provided by ____________.

209. A chemical reaction that transfers a phosphate group is called phosphorylation. ATP has ______ phosphate groups. ATP reacts with many other compounds and frequently transfers its third phosphate group to another molecule. This reaction is called ____________ and produces a phosphorylated biomolecule, as well as ______ (adenosine diphosphate).

210. A molecule that binds to a protein is called a ____________. Binding (or release) of a ligand usually triggers a slight reshaping of the protein, called a ________________. Proteins are ____________ (flexible or rigid) molecules that move as they work. Phosphorylation of a protein by ____________ frequently triggers a ________________ in the protein that helps the protein do its job.

211. Phosphorylation of a muscle protein called myosin triggers a ____________ in the myosin that is an important part of the muscle contraction process.

212. ____________ functions as an energy shuttle within the cell. It provides energy to reactions that require it by ________________.

Biomolecules: An Overview

The four major groups of biomolecules make up much of the structure of the cell and are responsible for most of its interesting activities. The diagram on the next page shows space-filling models of compounds from each of the four major groups. Look at these and try to imagine the molecules in action.

The enzyme is a typical example of a protein that has folded into its active, tertiary configuration. The electron cloud of this macromolecule is unique to this particular protein. You can imagine each protein as an irregular, three-dimensional jigsaw puzzle piece, each with its own special shape. Only a matching puzzle piece can bind to this protein. A complementary molecule has more than just a matching shape. The surfaces that fit each other must have chemical compositions that attract each other. Groups with opposite charge that can form an ionic bond would be a good example. When two molecules fit, bind to each other, and have the proper chemistry, something interesting happens. Proteins are responsive molecules that wiggle around when something attaches to them (or dissociates). These wiggles, or conformation changes, may facilitate a chemical reaction, transport a substance across a membrane, or enable a protein motor to chug along a track (probably also made of protein). The structure and function of a cell are largely a reflection of the proteins it contains.

The carbohydrate molecules, glucose and fructose, are isomers of each other. Though very similar chemically, they have slightly different shapes. Molecules interact through their electron clouds, and the shapes of these molecules are different enough to give them specificity within cells. A typical enzyme can discriminate between these two hexoses. For example, the enzymes that synthesize starch always bind glucose at the active site, but never fructose.

The shapes of the unsaturated and saturated fats are informative as well. It is obvious that the saturated fat molecules will pack more tightly, and that leads directly to a higher melting point. It takes energy to dislodge these molecules and get them moving independently, as they do in the liquid phase. The phospholipids provide a clear example of molecular self assembly. The chemical personalities of the two parts of the phospholipid molecule are so different that they don't associate with each other. The hydrophobic parts stay together (avoiding water at the same time). The hydrophilic parts stay in contact with each other and with water. The result is the phospholipid bilayer, a self-assembled structure with novel properties essential to life—the phospholipid bilayer is the basis of all biological membranes that enclose and partition cells.

Finally, consider the double helix, possibly the best-known shape of all the biomolecules. From the outside, the molecule looks repetitive, a spiralling structure that always looks the same. Most books look the same also, but open them up and they tell very different stories. They perform this feat by arranging 26 symbols, the letters of the English alphabet, into different sequences. How many words and stories are there? More than you can ever read! Think of DNA as a book. The spiral can come apart into two strands, and once dissociated, it is easy to see that the sequence of base pairs on the inside varies endlessly, each base with its own unique shape. The information coded in your DNA in the sequence of the four bases (a kind of chemical alphabet) is unique to you. No one else has ever had exactly the same sequence of DNA "letters" as you, and no one will ever have exactly this story again.

Since a gene acts by controlling the primary structure of a protein, it follows that each genetically distinct organism has a unique combination of proteins. And now we're back where we began, with proteins, a group that deserves its name of "first substance." Proteins make up much of the structure of the cell, and they are the workers that get things done. We see the results as traits, characteristics that we can observe, ultimately under genetic control. Taken together, the biomolecules are doing some amazing things, phenomena so complex that they account for the very special properties we associate with life.

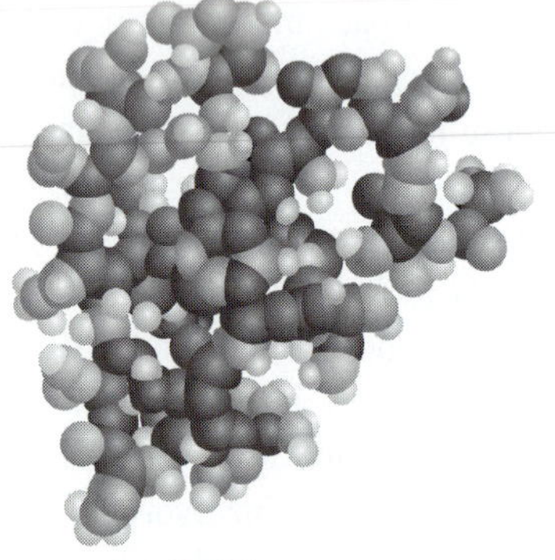

Protein (enzyme)

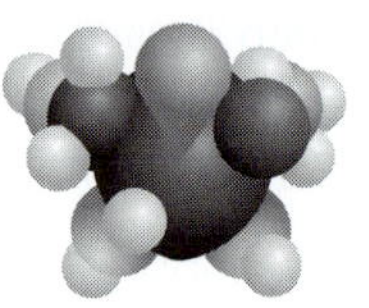

Glucose

Fructose

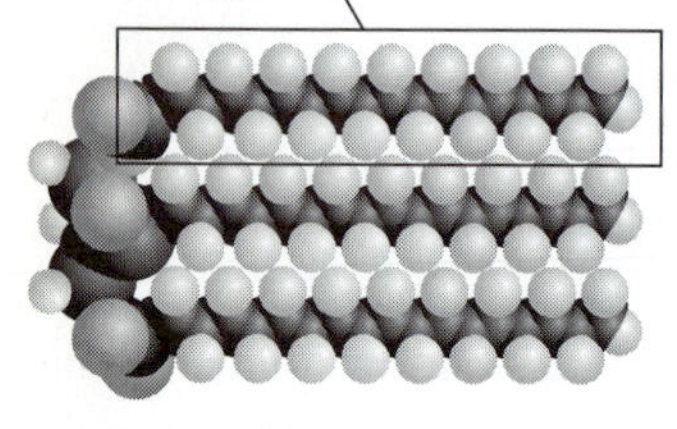

Saturated fat

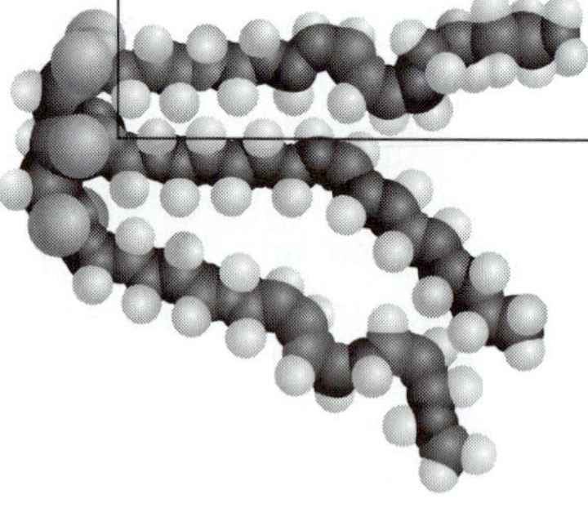

Unsaturated fat

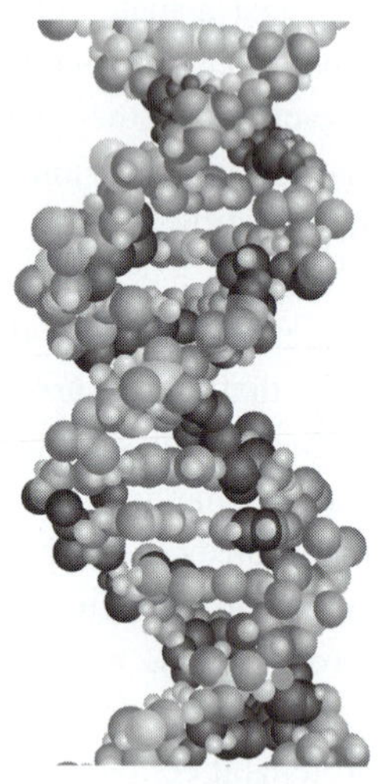

Nucleic acid (DNA)

Answers

1. decreases; water.

2. water.

3. organic; carbon.

4. inorganic; organic compounds.

5. Cellulose is organic. The question states that cellulose is a carbohydrate, and carbohydrates are one of the four major classes of organic compounds.

6. organic; carbon; inorganic.

7. polymer; monomers; water.

8. polymer; water molecules; oligomer.

9. monomers.

10. monomers; polymer; amino acids.

11. polymer; glucose.

12. amino acids; heteropolymer.

13.*Proteins are heteropolymers composed of 20 different kinds of monomers. They are far more variable than starch, which is a homopolymer.

In a heteropolymer such as protein, the monomers can be polymerized in different orders or sequences. Making necklaces using 20 different kinds of beads would be an appropriate analogy. Many different necklaces with strikingly different appearances could be made. Similarly, each unique sequence of amino acids forms a different protein, with a unique shape and function. There is only one type of monomer in a homopolymer such as starch. This is analogous to stringing necklaces using only one type of bead; all the necklaces are very similar. The size of different starch molecules can differ, and the degree of branching, if any, is also a variable. But sequence variation (different kinds of units in different orders) is not possible, so all starch molecules have very similar properties, and all function as energy reserves

14. more than one; nucleotides.

15. nucleotides; are not.

16. heteropolymer; nucleotides.

17. starch; heteropolymers; starch.

18. Proteins would have a greater variety of functions.

    Starches do not exhibit much variation in structure, because they are homopolymers and all function as energy reserves. Proteins exhibit great variety in their structure (discussed in the answer to question 13), and the different structures have different properties and different functions in the cell.

19. amino acids; amino acids.

20. fat or oil; polymer.

21. left; water; dehydration synthesis.

22. hydrolysis; monomers.

23. amino acids; nucleotides; glucose; homopolymer.

24. amino acids; dehydration synthesis.

25. many amino acids.

26. hydroxyl or alcohol (–OH); hydrogen.

27. dehydration synthesis; hydroxyl; hydrogen.

28. all

29. speed up; enzymes; protein.

30. more specific.

31. hemoglobin; hemoglobin.

32. are not; hydrocarbon; proteins that function as carrier molecules for lipids.

    The carrier proteins that transport fats and other lipids (including cholesterol) are called lipoproteins. These proteins are combined with varying amounts of lipid and have a range of densities. They are classed as: HDL—high-density lipoprotein, the so-called "good" cholesterol; LDL—low-density lipoprotein, the so-called "bad" cholesterol; and VLDL—very low-density lipoprotein.

33. structural proteins, such as those in skin and connective tissues.

34. Proteins are heteropolymers composed of 20 different kinds of monomers. In different proteins, the monomers are linked together in different orders or sequences. Each unique sequence of amino acids forms a different protein. Each different sequence causes the protein to fold into a unique shape. The unique shape of each protein suits it for a different function in the cell.

35. An amino group, $-NH_2$, and a carboxyl group, $-COOH$, are both present in amino acids.

36. Both the amino group, $-NH_2$, and the carboxyl group, $-COOH$, can ionize. In healthy cells and in most tissues, pH values are near neutral. At these pH values, some carboxyl groups ionize by releasing a hydrogen ion (H^+). The remainder of the carboxyl group ($-COO^-$) is then negatively charged. Some amine groups accept a hydrogen ion and become positively charged ($-NH_3^+$).

37. partially; hydrogen; hydrogen ions (H^+), buffer.

38. buffers; hydrogen ions.

39. One of the four groups attached to the central carbon atom is different. The other three groups attached to the carbon are the same in all amino acids and include a hydrogen, an amine group, and a carboxyl.

40. hydrophobic; hydrophilic.

41. yes. Yes, some amino acids would be more soluble than others in water.

    The ionized amine and carboxyl groups are hydrophilic, so all 20 amino acids dissolve in water to some extent. If the R group is hydrophobic, the amino acid as a whole tends to be less soluble. Amino acids with a polar or a charged R group tend to be more soluble in water, since all parts of the molecule are hydrophilic.

42. the carboxyl group; four; three.

43. amine; three; nitrogen; two.

44. peptide bond; nitrogen; dipeptide.

45. a tripeptide; water molecule; two.

46. oligopeptide; nine; nine.

    Sketch some short peptides to convince yourself of this. The number of linked amino acids, minus one, will be the number of peptide bonds that are present.

47. amino acids; hydrolysis; amino acids.

48. hydrolysis; amino acids; protein catalysts.

49. eleven; ten; oligopeptide or neuropeptide; amino.

50. 128; a gene; sequence of amino acids in the protein.

51. hydrolysis; enzyme; stable; hydrolysis.

52. the sequence of amino acids in it; a specific gene. (Each protein or polypeptide has a unique primary structure controlled by its own gene.)

53. different from; amino acids; gene (with the information for the collagen primary structure).

54. No, lysozyme would not form spontaneously. The machinery of the cell, including ribosomes, enzymes, and various small molecules, is required.

55. oxygen; unfold. Denatured hemoglobin will not transport oxygen. Denatured proteins do not have the proper shape and cannot function.

56. Raw egg white is transparent, viscous, and yellowish in color. Cooked egg white is solid and white.

57. denaturation; coagulation, precipitation, and changes in viscosity are all physical changes that can occur after denaturation.

    The type of denaturation that occurs when an egg is cooked is extreme and irreversible (the egg does not become raw again). In contrast, beating egg whites vigorously (the first step in preparing meringues) forms a frothy mixture that contains gently and partially denatured egg albumin. If the meringue is not baked, it will renature (at least partially) as time passes.

58. different.

59. shape; gene.

60. primary; the primary structure.

61. a gene; a mutation.

62. secondary; hydrogen; weak.

63. secondary; hydrogen bonds. In the β pleated sheet, hydrogen bonds form between polypeptide chains that are next to each other.

64. hydrophobic. Hydrophobic R groups tend to end up on the inside of the protein.

    These groups can not hydrogen bond to water and tend to be excluded from it. "Hydrophobic bonds," or "hydrophobic interactions," are terms used to describe the clustering of nonpolar molecules near each other due to their exclusion from water.

65. ionic.

66. water; hydrogen. Polar R groups tend to end up on the outer surface of the protein, because they can then hydrogen bond to water.

67. primary; secondary; tertiary; more than one.

68. secondary; tertiary; and quaternary (if present); primary; does not. Hydrogen bonds, ionic bonds; hydrophobic interactions, and disulfide linkages are broken when a protein denatures. These are the bonds or attractions that form between different R groups or between R groups and nearby water molecules.

69. peptide; hydrolysis; protease.

70. catalyst; substrates.

71. active sites; complementary; will not.

72. unfold. No, active sites are not present after denaturation, because the entire protein has unfolded. The active site is normally formed by a few amino acids that are close together in the properly folded protein, but are far apart in the denatured protein.

73. does not. Only an enzyme that is properly folded has an active site. The active site is essential to bind specifically to substrates, and substrates must bind to an enzyme before they can be converted to products.

74. complementary; are not; do not.

75. active site; active sites; have complementary shapes.

76. ligand; substrates; ligands.

77. conformation change; conformation change.

78. easier; faster.

79. ligand; conformation change.

80. binding (or release) of a ligand; unfolding; function; changes in the environment around the protein, such as shifts in pH or a significant rise in temperature.

81. protease; peptide; amino acids.

82. carbon, hydrogen, and oxygen; 1:2:1.

83. 12; six oxygens; isomers.

84. 10; 5; 0.

    Ribose is a simple sugar, and simple sugars contain only the elements carbon, hydrogen, and oxygen in a 1:2:1 ratio. The number of oxygen atoms will equal the number of carbon atoms in the sugar, and there will be twice as many hydrogen atoms.

85. one; sugar.

86. double sugar—the term refers to the entire molecule formed by linking two sugar molecules together.

87. many; glucose.

88. energy; is.

89. are. Concentrated sugar solutions are viscous and sticky.

90. polysaccharide; hydrolyzed.

91. three; five; six.

92. sugar; carbon atoms.

93. two; four; five. The number of alcohol groups is equal to the number of carbon atoms in the sugar, minus one. Every carbon atom except one has a hydroxyl group attached to it.

94. aldehyde; ketones.

95. polar; do.

96. Glyceraldehyde, glucose, and ribose all have an aldehyde group. Fructose contains a ketone group.

97. No, carbohydrates should contain only the elements carbon, hydrogen, and oxygen.

98. ring. α glucose and β glucose.

99. ring. The position of one of the five alcohol (or hydroxyl) groups present in glucose is different in the α form of glucose, as compared to the β form. In α glucose, the hydroxyl group on the far right is directed down, as is the hydroxyl on the opposite end. In β glucose, the hydroxyl groups on the opposite ends of the molecule point in different directions. Each form of glucose is somewhat different in shape

100. are; α; dehydration synthesis; α; β.

101. β; dehydration synthesis; β; α.

102. glucose; enzyme; glucose. (The enzyme that catalyzes the hydrolysis of starch is called amylase.)

103. starch; hydrolyze; sugar (or glucose). Starch is used as an energy reserve

When excess sugar is present in the cell, it is polymerized to form starch (and water) and stored. When sugar is needed again, the starch can be hydrolyzed. Starch can be stored in solid form, in granules, and does not cause the problems that concentrated sugar solutions would have.

104. fructose; glycosidic; disaccharide.

105. dehydration synthesis. A catalyst.

Virtually every reaction in living organisms is catalyzed by a specific enzyme that greatly increases the rate of that reaction. Enzymes increase reaction rates so much that a reaction occurring at a barely detectable rate without an enzyme, can occur rapidly with its enzyme. Enzymes are literally responsible for metabolism, the chemical activity of life.

106. glucose and fructose; hydrolysis; water. The specific enzyme that catalyzes this reaction is called sucrase. With sucrase, the hydrolysis reaction occurs rapidly.

107. disaccharide; enzyme; protein; disaccharide; maltase.

The fact that there is a unique enzyme for the hydrolysis of each disaccharide illustrates the great specificity of enzymes. The shapes of maltose, lactose, and sucrose are somewhat different. The active site of an enzyme discriminates among them and binds (and hydrolyses) only the one with a complementary shape at its active site.

108. polysaccharide; glycosidic; glucose.

109. polysaccharide; sugar and energy reserve.

110. sugar; hydrolyze (or digest); cannot. No, cellulose is not a calorie source for humans.

Since we cannot hydrolyze cellulose, sugar is not released and absorbed from it. Large molecules such as cellulose cannot be absorbed. Since it is not digested, cellulose passes right through the digestive tract and is an important component of dietary fiber, which promotes normal motility of the digestive tract.

111. glucose; glucose; glucose; glucose; homopolymers.

112. heteropolymers; fewer; they are not as structurally diverse. Since function is always directly related to structure, lack of variety in the structures leads to lack of variety in the biological roles.

113. α glucose; glucose.

114. They differ in the degree of branching that is present. Glycogen is extensively branched, and amylose is not branched.

115. plant; cell walls. Cellulose is a polymer of β glucose.

The shape of β glucose differs from α glucose. When β glucose polymerizes, the cellulose linkage is also different in shape, as compared to a glycosidic linkage. Finally, the polymer as a whole is different in shape. As a reflection of this, cellulose has different properties. Fibrils of cellulose interact via large numbers of hydrogen bonds. Cellulose is strong, fibrous, and difficult to hydrolyze, and it is an excellent structural material.

116. cellulases; hydrolyze.

117. Humans do not produce cellulase themselves and also do not have microorganisms in the digestive tract that can digest cellulose.

118. hydrocarbon; hydrophobic.

119. energy; energy.

120. structural.

121. are not.

122. hydrocarbon; not; pain and smooth muscle contraction.

123. energy; hydrocarbon; energy; 1.

124. Fats store more energy per unit weight (9 calories per gram) than any other biomolecule. In contrast, carbohydrates and proteins store only about 4 calories per gram. Molecules with a high proportion of hydrogen, like the lipids, tend to be lightweight and high in stored energy.

125. three; eight; three; hydroxyl—glycerol can be described as a trialcohol, an organic compound with three hydroxyl groups within it.

126. hydrocarbon; carboxyl; hydrophobic; hydrogen.

127. Fatty is a good description of the hydrophobic part of the molecule that is composed of hydrocarbon. Acid describes the carboxyl group, a weak organic acid that is partially dissociated.

128. hydrogen; stearate; oleate.

129. 14; 18; 18; 18; even.

130. Stearic acid, saturated: $CH_3(CH_2)_{16}COOH$;
Oleic acid, monounsaturated: $CH_3(CH_2)_7CH = CH(CH_2)_7COOH$;
Linolenic acid, polyunsaturated: $CH_3CH_2CH = CHCH_2CH = CHCH_2CH = CH(CH_2)_7COOH$;

131. single bonds; two; single bonds; two; double bond.

132. saturated; single bonds; hydrogen.

133. less; double; four; one.

134. The hydrocarbon portion of a saturated fatty acid is a straight cylinder. Unsaturated fatty acids have bends at the position of each double bond.

135. hydrogen; dehydrogenation.

136. hydrogenation; hydrogen.

137. dehydrogenation; hydrogen; double.

138. hydrogenation; reduction.

139. three; one; three.

140. hydrolysis; three fatty acids and a glycerol molecule.

141. dehydration synthesis; increase.

142. hydrolyzing; fatty acids; as an energy source or fuel.

143. a stored energy reserve.

144. unsaturated; saturated.

145. do not; are. Butter, lard, and tallow (from sheep) are all solid at room temperature and contain mostly saturated fats.

146. does; is not. Corn oil, olive oil, peanut oil, and sesame oil are all examples of fats that are liquid at room temperature. They contain mostly unsaturated fats.

147. Fatty acids have a carboxyl group that can ionize and develop a charge. When the fatty acids are bonded to glycerol, ester linkages form. Ester linkages do not ionize and become charged; hence the name neutral fats. Triglyceride is also a good description of a fat, because the fat contains three fatty acids bonded to glycerol.

148. linear or straight; bent and crooked. The crooked shapes of unsaturated fatty acids do not pack together as closely as do saturated fats. Molecules that are farther apart cannot form weak attractions to each other as effectively as molecules that are closer together. Therefore, unsaturated fatty acids tend to have a lower melting point than those that are saturated (and therefore packed closer together). The diagrams at the end of the chapter illustrate this.

149. Fats that are solid at room temperature, such as butter, lard, hard margarines, and the like.

150. The more saturated the fat, the higher the melting point.

151. liquid; unsaturated; solid. The unsaturated fats in the corn oil have been partially hydrogenated.

Hydrogen has been added, breaking some of the double bonds between carbon atoms in the backbone. This produces a fat with a higher melting point. Unfortunately, the healthful benefits of the polyunsaturated corn oil are lost, along with the double bonds.

152. saturated; higher.

153. plants; oils; seeds. The oils store a lot of energy in a minimal weight.

The embryonic plant depends on food that is stored in the seed until it grows enough to break through the soil, form its first leaves, and begin photosynthesis.

154. two; ester; phosphate; H^+ (hydrogen ion).

155. hydrocarbon; nonpolar; hydrophilic.

156. circle; a phosphate group and a nitrogenous base.

157. two lines; the hydrocarbon chains of the fatty acids.

158. hydrophilic; nonpolar.

159. The charged part of the phospholipid is likely to be next to water. Structures that are partially or completely charged are hydrophilic, because they can form hydrogen bonds with water.

160. two; hydrophobic; hydrophilic;

161. hydrophobic; hydrophilic

162. The hydrocarbon chains of the fatty acids are hydrophobic, because they are nonpolar. They tend to be excluded from water, because they cannot interact with it to form hydrogen bonds. The tendency of hydrophobic groups to aggregate (called hydrophobic interactions) results mostly from the tendency of water to exclude these structures. Hydrocarbons attract each other only very weakly.

163. Groups that are hydrophilic are attracted to water and to each other. They can form ionic bonds and hydrogen bonds when they are correctly positioned. This stabilizes the bilayer.

164. phospholipid bilayers; self assemble.

165. phospholipid bilayer. The glucose transporter is a membrane-bound protein; that is, it is embedded in the phospholipid bilayer of the cell's outer membrane (the plasma membrane). It spans the membrane and is therefore effective as a protein that transports glucose from the outside to the inside of the cell.

166. four; hydrocarbon. Different steroids may have different numbers of double bonds within the ring structure. They also have different groups attached to the rings.

167. steroid; hydrocarbon; hydrophobic. Since cholesterol is hydrophobic, it will be found in the hydrophobic portion of the phospholipid bilayer. This region also contains the hydrocarbon "tails" of the fatty acids.

168. sex hormone; protein.

169. four hydrocarbon rings. They would have different groups attached to the rings and might have different numbers of double bonds at various locations in the rings.

170. cholesterol; sex hormone.

171. DNA and RNA; nucleotides; more than one.

172. larger than.

173. DNA; RNA.

174. gene; DNA; amino acid sequence—the primary structure.

175. DNA. The nucleotides are in different sequences, or orders, in different DNA molecules.

176. nucleotides; amino acids; DNA; amino acids.

177. shape; function; denaturation; sequence; function. The hemoglobin that is produced in a person who has sickle cell anemia has a different primary structure than normal hemoglobin. Its shape and function are therefore changed. Changes in the function of the hemoglobin produce the disease symptoms of sickle cell anemia.

178. five; ribose; deoxyribose. Deoxyribose contains one less oxygen. Rather than having an alcohol group at the second carbon, it has only a hydrogen.

179. nitrogen; can accept a hydrogen ion (H^+). Adenine, guanine, cytosine, thymine, and uracil are nitrogenous bases.

180. adenine, guanine, and cytosine; thymine; uracil.

181. adenine — A guanine — G cytosine — C thymine — T uracil — U

182. either ribose or deoxyribose; A, G, C, and T; A, G, C, and U.

183. nitrogenous bases; primary structure; bases.

184. polypeptide (or protein); nitrogenous bases.

185. polymers; dehydration synthesis; water.

186. phosphate.

187. The nitrogenous bases are attached to the sugar (ribose or deoxyribose) and are on the side of the sugar-phosphate backbone.

188. bases.

189. covalent.

190. nucleotides; provide a specific enzyme to hydrolyze the DNA.

191. -ase; hydrolyses.

192. DNA; RNase.

193. covalent; hydrogen.

194. adenine (A)

195. guanine (G)

196. hydrogen; complementary; G; T or A or another C.

197. two; three.

198. double helix. The base pairs are inside the double helix, rather like the stairs in a spiral staircase. The sugar-phosphate backbones are on the outside, like the banisters on a spiral staircase.

199. polynucleotide; four; heteropolymer.

200. single stranded. RNA contains the sugar ribose, rather than deoxyribose, and it replaces the base thymine with uracil.

201. single stranded.

202. G; uracil (U).

203. GGCAATGCT-; 23; broken.

204. GGCAAUGCU-. It is the same, except that U is present instead of T.

205. ATP is like the nucleotide that contains adenine, with two additional phosphate groups present.

206. ATP

207. ATP

208. ATP

209. three; phosphorylation; ADP.

210. ligand; conformation change; flexible; ATP; conformation change.

211. conformation change.

212. ATP; phosphorylation.

Chapter Test: Biomolecules

The questions in this chapter test evaluate your mastery of all the objectives for this unit. Take this test without looking up material in the chapter. After you have completed the test, you can check your work by using the answer key located at the end of the test.

1. The most prevalent compound in most living cells is ________________.

2. The four groups of biomolecules include ____
____

3. Small molecules called __________ can be linked together by a type of reaction called __________ to form __________.

4. Define the terms listed below:

   macromolecule ____

   heteropolymer ____

5. Fill in the missing information in the table below.

Group	Monomers	Polymers
Proteins		Peptides and proteins
Carbohydrates	Sugars	
Nucleic acids		
Lipids	Fatty acids and glycerol	

6. Biological catalysts are composed mostly of __________ and are called __________.

7. List three different functions performed by proteins in a typical cell.

____

8. What two functional groups are always present in amino acids? ____________________________

9. A structure composed of six linked amino acids could be called a ___________ or an ______________________. This structure would contain ___________ peptide bonds.

10. An enzyme that catalyzes the hydrolysis of proteins is called a ___________; an enzyme that catalyzes the hydrolysis of DNA is called ___________; and an enzyme that catalyzes the hydrolysis of sucrose is called ___________.

11. What is the primary structure of a protein ? ____
____

12. What bonds are broken when a protein is hydrolyzed? ___________ If the hydrolysis process is complete, you would then have a mixture of ____.

13. What bonds are broken when a protein is denatured? ________________________________
____
Describe a denatured protein structurally and functionally. ______________________________
____

14. What is a conformation change, and what causes a conformation change to occur?
____
____

15. Hemoglobin is a ___________ that exhibits the ___________ level of structure, because it contains four polypeptide chains within one functional hemoglobin unit. Hemoglobin functions as a carrier molecule that transports ___________ within many animals.

16. A ___________ is a permanent, heritable change in the ___________ that makes up the structure of a gene. A mutation can cause a change in the ___________ structure of a protein. This can change the shape and ___________ of the protein. An alteration in the function of a protein may

produce a noticeable change in a trait in a cell and an entire organism. Give an example of such a mutation, and explain how the mutation changes the organism.

____

____

17. A compound with the formula $C_5H_{10}O_5$ would most likely be a ____________, because the elements carbon, hydrogen, and oxygen are present in a ____________ ratio.

18. A trisaccharide ____________ (could or could not) be hydrolyzed and would yield ____________ sugar molecules when the reaction was completed.

19. Simple carbohydrates are ____________ molecules that are soluble in water and can therefore be classed as ____________.

20. Polymerization of the ______ form of glucose produces starch. Polymerization of β glucose produces ____________.

21. The major reaction of digestion is __________________.

22. Cellulose is synthesized by ____________ and is used to construct ____________. Amylose and glycogen function mainly as ____________.

23. One class of biomolecules is hydrophobic, because it is composed mostly of ____________. These biomolecules are called ____________.

24. A saturated fatty acid contains more ____________ than an unsaturated fatty acid. Unsaturated fatty acids contain ____________ bonds between some carbon atoms, whereas saturated fatty acids do not.

25. Lipids that function as energy reserves are ____________; lipids that function as sex hormones are examples of ____________; and lipids that form the basic structure of cellular membranes are ____________.

26. A nucleotide contains __________________, ________________________, and one of four nitrogenous ____________.

27. Genes are composed of ____________, a nucleic acid. RNA is also a nucleic acid that is involved in several aspects of __________________.

28. How do genes code information? ____

29. How do genes control traits within cells? ____

____

30 The double helix describes the shape of a ____________ molecule.

31. What is ATP and what is its role within the cell? ____

32. List three ways that most RNA molecules differ from a typical DNA molecule.

____

33. Within DNA, A pairs with ____________, and G pairs with ____________.

Answers

1. water

2. proteins, carbohydrates, lipids, and nucleic acids. (These answers can be in any order.)

3. monomers; dehydration synthesis; polymers.

4. A macromolecule is a large molecule, usually a polymer. It is formed by linking many small subunit molecules, called monomers, together. A typical monomer contains hundreds or even thousands of monomers, linked together by covalent bonds.

   A heteropolymer is a polymer that contains more than one kind of monomer within it. Different sequences of the monomers then form polymers with different structures and properties.

5.

Group	Monomers	Polymers
Proteins	Amino acids	Peptides and proteins
Carbohydrates	Sugars	Starch and cellulose
Nucleic acids	Nucleotides	DNA and RNA
Lipids	Fatty acids and glycerol	Fats and oils

6. protein; enzymes.

7. Various proteins function as: enzymes, carrier molecules, structural molecules, defensive proteins, communication between cells, and in movement.

8. amino group and a carboxyl group.

9. hexapeptide; oligopeptide; five.

10. protease; DNase; sucrase.

11. The primary structure of a protein is the sequence of amino acids within it.

12. peptide bonds; the amino acids that were present in the protein.

13. hydrogen bonds, ionic bonds, and disulfide linkages. All the bonds between amino acid side groups, with each other and with water, break during denaturation. Peptide bonds do not break.

    A denatured protein is unfolded and does not have its normal shape. A denatured protein does not function, because the normal protein shape is essential for normal function.

14. A conformation change is a slight reshaping or wiggling of the protein. It can be triggered by binding of a ligand, or by release of a ligand. Conformation changes are often important to the normal function of the protein.

15. protein; quaternary; oxygen.

16. mutation; DNA; primary; function. Sickle cell anemia (SCA) is a genetic disease that originated when a gene for globin (one of the components of hemoglobin) had a specific mutation.

    The mutated gene codes for a slightly different globin which has one amino acid that is different from the common, "normal" type of hemoglobin. This altered primary structure leads to a small, but significant, change in shape of the folded hemoglobin. The SCA form of hemoglobin has somewhat unique properties. It tends to crystallize when oxygen levels are low. The crystallized SCA hemoglobin distorts, or sickles, the red blood cell that contains it. Sickled cells tend to become stuck in the small capillary blood vessels, impeding local circulation and causing pain. Sickled cells tend to lyse, leading to anemia. Thus, an altered gene produces a changed protein that functions differently, leading to the traits associated with SCA.

17. sugar; 1:2:1

18. could; 3

19. polar; hydrophilic.

20. α; cellulose.

21. hydrolysis.

22. plants; cell walls; fuels—they are reserves of glucose that can be released by hydrolysis when needed.

23. hydrocarbon; lipids.

24. hydrogen; double.

25. fats and oils; steroids; phospholipids.

26. phosphate; ribose (in RNA) or deoxyribose (in DNA); bases.

27. DNA; gene expression.

28. Genes code information in the sequences of the nitrogenous bases, much as an alphabet codes information in different sequences of letters.

29. Genes control the primary structure of a protein. The primary structure of a protein determines its shape and therefore its function. Proteins are the workhorses of the cell. If a protein changes, the effects will often be seen in the structure or function of the cell and the organism as a detectable trait (See the answer to question 16 for an example.)

30. DNA

31. ATP stands for adenosine triphosphate, and it functions as an energy shuttle within the cell.

32. RNA is usually single stranded, it contains ribose rather than deoxyribose, and it contains uracil in place of thymine.

33. T (thymine); C (cytosine). The base pairing rules result from the fact that only bases that are complementary in shape can pair with each other. Other combinations do not fit.

PERIODIC CHART

6 Carbon **C** 12.01	Atomic number Name Symbol Atomic mass

☐ Metals
☐ Metalloids
☐ Nonmetals

Transition metals (groups 3–9)

Period	1 Group IA	2 IIA	3 IIIB	4 IVB	5 VB	6 VIB	7 VIIB	8 VIIIB	9 VIIIB
1	1 Hydrogen **H** 1.008								
2	3 Lithium **Li** 6.94	4 Beryllium **Be** 9.01							
3	11 Sodium **Na** 22.99	12 Magnesium **Mg** 24.31							
4	19 Potassium **K** 39.10	20 Calcium **Ca** 40.08	21 Scandium **Sc** 44.96	22 Titanium **Ti** 47.90	23 Vanadium **V** 50.94	24 Chromium **Cr** 52.00	25 Manganese **Mn** 54.94	26 Iron **Fe** 55.85	27 Cobalt **Co** 58.93
5	37 Rubidium **Rb** 85.47	38 Strontium **Sr** 87.62	39 Yttrium **Y** 88.91	40 Zirconium **Zr** 91.22	41 Niobium **Nb** 92.91	42 Molyb-denum **Mo** 95.94	43 Technetium **Tc** (98)	44 Ruthenium **Ru** 101.07	45 Rhodium **Rh** 102.90
6	55 Cesium **Cs** 132.91	56 Barium **Ba** 137.34	* 57 Lanthanum **La** 138.91	72 Hafnium **Hf** 178.49	73 Tantalum **Ta** 180.95	74 Tungsten **W** 183.85	75 Rhenium **Re** 186.21	76 Osmium **Os** 190.2	77 Iridium **Ir** 192.22
7	87 Francium **Fr** (223)	88 Radium **Ra** 226.03	† 89 Actinium **Ac** 227.03	104 Unnil-quadium **Unq** (261)	105 Unnil-pentium **Unp** (262)	106 Unnil-hexium **Unh** (263)	107 Unnil-septium **Uns** (262)	108 Unnil-octium **Uno** (265)	109 Unnil-ennium **Une** (266)

*Lanthanides	58 Cerium **Ce** 140.12	59 Praseo-dymium **Pr** 140.91	60 Neo-dymium **Nd** 144.24	61 Prome-thium **Pm** (145)	62 Samarium **Sm** 150.36	63 Europium **Eu** 151.96	64 Gadolin-ium **Gd** 157.25
†Actinides	90 Thorium **Th** 232.04	91 Protac-tinium **Pa** 231.04	92 Uranium **U** 238.03	93 Neptunium **Np** (237)	94 Plutonium **Pu** (244)	95 Americium **Am** (243)	96 Curium **Cm** (247)

OF THE ELEMENTS

10	11 IB	12 IIB	13 IIIA	14 IVA	15 VA	16 VIA	17 VIIA	18 VIIIA
								2 Helium **He** 4.00
			5 Boron **B** 10.81	6 Carbon **C** 12.01	7 Nitrogen **N** 14.01	8 Oxygen **O** 16.00	9 Fluorine **F** 19.00	10 Neon **Ne** 20.18
			13 Aluminum **Al** 26.98	14 Silicon **Si** 28.09	15 Phos-phorus **P** 30.97	16 Sulfur **S** 32.06	17 Chlorine **Cl** 35.45	18 Argon **Ar** 39.95
28 Nickel **Ni** 58.69	29 Copper **Cu** 63.54	30 Zinc **Zn** 65.37	31 Gallium **Ga** 69.72	32 Germa-nium **Ge** 72.61	33 Arsenic **As** 74.92	34 Selenium **Se** 78.96	35 Bromine **Br** 79.91	36 Krypton **Kr** 83.80
46 Palladium **Pd** 106.42	47 Silver **Ag** 107.87	48 Cadmium **Cd** 112.41	49 Indium **In** 114.82	50 Tin **Sn** 118.69	51 Antimony **Sb** 121.75	52 Tellurium **Te** 127.60	53 Iodine **I** 126.90	54 Xenon **Xe** 131.29
78 Platinum **Pt** 195.08	79 Gold **Au** 196.97	80 Mercury **Hg** 200.59	81 Thallium **Tl** 204.37	82 Lead **Pb** 207.19	83 Bismuth **Bi** 208.98	84 Polonium **Po** (209)	85 Astatine **At** (210)	86 Radon **Rn** (222)

65 Terbium **Tb** 158.92	66 Dys-prosium **Dy** 162.50	67 Holmium **Ho** 164.93	68 Erbium **Er** 167.26	69 Thulium **Tm** 168.93	70 Ytterbium **Yb** 173.04	71 Lutetium **Lu** 174.97
97 Berkelium **Bk** (247)	98 Califor-nium **Cf** (251)	99 Einstei-nium **Es** (252)	100 Fermium **Fm** (257)	101 Mende-levium **Md** (258)	102 Nobelium **No** (259)	103 Lawren-cium **Lr** (260)